Ergebnisse der Physiologie · Reviews of Physiology

Ergebnisse der Physiologie
Biologischen Chemie und experimentellen Pharmakologie

Reviews of Physiology
Biochemistry and Experimental Pharmacology

63

Herausgeber / Editors

E. Helmreich, Würzburg · H. Holzer, Freiburg · R. Jung, Freiburg
K. Kramer, München · O. Krayer, Boston
F. Lynen, München · P. A. Miescher, Genf · W. D. M. Paton, Oxford
H. Rasmussen, Philadelphia · A. E. Renold, Genf
U. Trendelenburg, Würzburg · H. H. Weber, Heidelberg

With 79 Figures

Springer-Verlag Berlin Heidelberg GmbH 1971

ISBN 978-3-662-31040-3   ISBN 978-3-540-36443-6 (eBook)
DOI 10.1007/978-3-540-36443-6

Library of Congress Catalog Card Number 62-37142.

Universitätsdruckerei H. Stürtz AG, Würzburg

Inhalt

Mitarbeiter

Schaper, W., Priv.-Doz. Dr., Department of Cardiovascular Research, Janssen Pharmaceutica, Koninklijeke Laan 17, B-2340 Zie Beerse

Schild, H. O., Prof. Dr., Department of Pharmacology, University College London, Gower Street, London W.C.1/Great Britain

Schmid-Schönbein, H., Priv.-Doz. Dr., Physiologisches Institut der Universität, D-8000 München 15, Pettenkoferstr. 12

Schmidt, R. F., Prof. Dr. Dr., Physiologisches Institut der Universität, D-2300 Kiel, Olshausenstr. 40/60

Wells, R. E., Dr., Department of Medicine, Peter Bent Brigham Hospital and Harvard Medical School, 721 Huntington Avenue, Boston, Mass. 02115/USA

Henry Hallett Dale, 1875—1968

H. O. SCHILD

"Favoured with opportunities such as few of my own generation enjoyed, I have been able to make scientific research, on problems mainly of my own choosing, the central activity of my working life". These introductory words to *Adventures in Physiology*, a selection of DALE's scientific writings published in 1953, emphasize the central role which scientific research has played in his life, but they can hardly convey the feeling of passionate involvement in scientific work which emanated from his personality. His conversation, even in old age, was outstandingly lucid and informative. Talking to him on a scientific subject was like wandering through a sunlit landscape in which every detail is distinctly delineated; the clarity, intensity and confidence of his thought adding to its impact.

DALE was one of the originators of modern physiology and almost certainly the most influential pharmacologist since PAUL EHRLICH. His work has shaped much of contemporary medicine and the power and fruitfulness of his thought was such that we now tend to think within the framework of his ideas. DALE remarked, in a paper on EHRLICH, whom he greatly admired, that a predominant line of interest and habit of thought could be traced in all his work. This applies equally to DALE himself; in his case the idea of autoregulation in the body, or as he himself called it, autopharmacology, provided the connecting link.

DALE's scientific development was extraordinarily harmonious and seemed to depend on certain dominating ideas which he developed more or less throughout his life. The following is a brief account of his main scientific achievements.

Scientific Work
Ergot and the Dual Action of Adrenaline

DALE's work, carried out in 1909–1912, has been the source of most subsequent developments in this field, including some of the most recent. It provides a good illustration of his approach and of his capacity to extract fundamental notions out of scattered and often accidental experimental observations. The following is his own (abbreviated) account of his first important experimental discovery. "When I accepted the appointment (to the Wellcome Physiological Research Laboratories) Mr. WELLCOME said to me that, when I could find an opportunity for it without interfering with plans of my own, it would give him special satisfaction if I would make an attempt to clear up the problem of ergot, then in a state of obvious confusion. I was frankly not at all attracted by the prospect of making my first excursion into pharmacological research on the ergot morass, but I found GEORGE BARGER, a former Cambridge colleague, and a very able and enterprising chemist, working in the Wellcome laboratories, and BARGER had already prepared from ergot a number of active substances. So I began by testing their effects on the blood pressure of the cat. All of these compounds showed an initial pressor action but that in itself would not have added greatly to earlier knowledge. By one of my greatest strokes of good fortune, however, this work was to give me an opportunity of making a mistake of my own—a really shocking 'howler'. I was finishing one of these experiments on a spinal cat to which I had given successive doses of one of BARGER's ergot preparations when a sample of dried suprarenal gland substance was delivered to me from the Burroughs Wellcome factory, with a request that I would test it for the presence of adrenaline. Successive injections of the extract elicited, to my surprise, only falls in arterial pressure, and with the confidence of inexperience I condemned the sample without hesitation. And then by another and almost incredibly fortunate coincidence, the same sequence of events was repeated in detail a week later. Again I was finishing an experiment on a cat heavily dosed with an ergot preparation when a sample of suprarenal gland substance was delivered for testing. The result was the same as on the earlier occasion and by reference to my notes now raised the question whether the cat's response to adrenaline might have been so altered by ergot that the normal pressor action had been reversed."

DALE's following paper deals with a detailed analysis of the interaction between ergot and adrenaline. He concluded that ergot paralysed the *motor functions* of the sympathetic and of adrenaline, leaving *inhibitory functions* unaffected. Although DALE was aware of LANGLEY's postulated *receptor substance* on which adrenaline supposedly acted, he makes no reference to receptors either in this or in later papers. He suggests instead that sympathetic nerves might contain some fibres connected to purely inhibitory and others to purely motor *myoneural junctions*, or alternatively that myoneural junctions may themselves be composite, containing both motor and inhibitory elements. DALE also observed that the actions of adrenaline on the heart were much more difficult to antagonise by ergot than its actions on blood vessels, thus anticipating in this early paper the modern subdivision into adrenergic α- and β-receptors.

Sympathomimetic Amines and Noradrenaline

DALE and BARGER's study of over fifty sympathomimetic amines is a comprehensive investigation of the relation between chemical constitution and pharmacological action. They introduced the term *sympathomimetic* to denote a range of compounds which "simulate the effects of sympathetic nerves with varying intensity and varying precision." It is characteristic of DALE's way of thinking to have adopted this rather open definition; he was generally averse to rigid formulations, preferring to use a more pragmatic approach. Their main conclusions were that the more substances resembled adrenaline in structure the more active they became, and that all typical sympathomimetic compounds were primary or secondary amines whilst tertiary and quaternary bases had nicotine-like actions.

Retrospectively, perhaps the most interesting aspect of their paper is its discussion of noradrenaline. DALE investigated noradrenaline showing that it had greater stimulant and less inhibitory activity than adrenaline and he points out in the discussion that the actions of noradrenaline correspond more closely to sympathetic stimulation than those of adrenaline. He explicitly rejects ELLIOTT's earlier suggestion that the similarity between the effects of adrenaline and sympathetic nerves may be due to the latter liberating adrenaline at nerve endings, giving as his main reason the lack of parallelism between adrenaline and sympathetic stimulation; but he fails to conclude that noradrenaline might occur as such in the body and be a better candidate than adrenaline of ELLIOTT's postulated transmitter substance. DALE himself marvelled later at his failure to "jump to the truth". In this connection I can recall when working with DALE in 1933 and finding a consistent discrepancy between physiological and colorimetric estimates of the adrenaline content of suprarenal glands that it did not seem to occur to either of us that the discrepancy might be due to the adrenal gland containing noradrenaline as well as adrenaline. It remained for the later work by HOLTZ and VON EULER to establish clearly the occurrence and function of noradrenaline in the body.

Histamine

DALE had a lifelong attachment to histamine, which he discovered (with LAIDLAW) in ergot extracts and whose pharmacological actions he first described. He used histamine as a tool for a deep-going analysis of capillary function under normal and pathological circumstances. In the course of this analysis he introduced the concept of a "locally acting chemical stimulant", distinguishing between true hormones in the Bayliss and Starling sense and other physiologically active substances whose effects were restricted to the immediate neighbourhood of their liberation. The problem of capillary tone had become of central interest through the work of AUGUST KROGH, and DALE thought that substances with "histamine-like" action played an important part in its regulation. He argued that the capillary epithelium must be constantly exposed to products with vasodilator activity which arose in tissues and whose production was accelerated by excitatory or injurious influences.

He considered that the simultaneous occurrence throughout the body of this type of capillary dilator activity would bring about a condition of irremediable circulatory collapse, or shock, through the stagnation of practically all the blood in the peripheral vessels. DALE and LAIDLAW showed that histamine could indeed produce effects akin to traumatic shock. The central feature of injecting a large dose of histamine was a loss of blood volume due to loss of tone by the capillaries, accompanied by an excessive permeability of blood vessels allowing the escape of plasma. Histamine shock was aggravated by anaesthesia.

DALE was, however, reluctant to claim a physiological role for histamine and at no point does he suggest that histamine is a physiological capillary regulator. This cautious attitude was due partly to lack of evidence that histamine was a normal constituent of tissues, but even at a later stage, after his own laboratory had shown that histamine could be extracted from tissues, he would go no further than to admit that it might be involved as a capillary dilator in reactions to superficial injury such as LEWIS's triple response.

In the sense that he failed to establish a physiological role for it, DALE's work on histamine was perhaps not as far-reaching in its consequences as his work on acetylcholine. Nevertheless his papers on histamine are well worth reading, not only because they describe its pharmacological actions almost completely (except the action on gastric secretion), but as examples of his scientific style, powers of physiological analysis, mastery of technique and wide pathological and medical range. They also illustrate the penetrating quality of his discussion and his care in drawing conclusions from experiments.

Anaphylaxis

DALE's work on anaphylaxis consists of a small number of papers which had a lasting influence on immunology. He entered the field by way of another "most fortunate" accident when discovering that an isolated uterus from a guinea-pig which had previously been accidentally injected with horse serum gave a maximal contraction with a further dose of horse serum. Although the same reaction was discovered independently by SCHULTZ in Washington, and their joint names are rightly attached to it, it was DALE's analysis of the underlying mechanism which won the day.

DALE's experiments had a quality of diagrammatic simplicity. He demonstrated that isolated sensitized preparations gave rise to a highly specific hypersensitivity reaction capable of distinguishing between hen and duck albumen; that serum was not involved in the reaction which would also occur after prolonged perfusion with Ringer solution; that cellular fixation of antibody was an essential feature of the reaction as shown by the latent period required for passive sensitization *in vitro*. Of great theoretical importance was DALE's clarification of the *desensitization* phenomenon which he clearly distinguished from *immunity*. He considered desensitization to be the lack of reaction of sensitized tissue to a second dose of antigen; immunity, the situation exemplified in experiments with KELLAWAY in which sensitized tissue was protected from the effect of antigen by keeping it immersed in a solution containing excess antibody.

DALE's papers were written at the height of the controversy between proponents of the cellular and humoral theories of anaphylaxis and it would probably be agreed today that his strictures on the anaphylatoxin theory were too harsh. Nevertheless his picture of the cellular basis of anaphylaxis has remained essentially intact, protected by the many careful qualifications with which he expressed his views. It is interesting that for a long time DALE was unwilling to accept a histamine theory of anaphylaxis, although he recognised from the start the similarities between effects of histamine and anaphylaxis. At one stage he suggested that the anaphylactic reaction might serve as a model for the action of histamine rather than the reverse, considering the reaction between cell-bound antibody and antigen to be the only known model of a drug-receptor interaction. Nevertheless he ended by proposing a histamine theory of anaphylaxis (in his Dohme lectures) several years before the histamine theory was experimentally verified through the elegant work of FELDBERG and DRAGSTEDT.

Dale's Work on Acetylcholine

Acetylcholine was synthesized in 1894 by NOTHNAGEL. In 1906 HUNT and TAVEAU discovered that it had a powerful hypotensive effect when injected into the mammalian circulation, and they already speculated on the possibility that it might be a natural constituent of the body, but this point was not proved until much later when DALE and DUDLEY extracted acetylcholine from horse spleen. DALE became interested in acetylcholine, as in histamine, by finding it (with EWINS) in extracts of ergot. He carried out a systematic investigation of acetylcholine in 1914, which revealed a striking parallelism between its actions and parasympathomimetic nerve stimulation. He found that the vasodilator and parasympathetic actions of acetylcholine were readily antagonised by atropine, but that after atropine the injection of larger doses of acetylcholine produced a nicotine-like stimulation of ganglion cells and the adrenal medulla. This finding induced him to subdivide the actions of acetylcholine into "muscarine" and "nicotine" actions with far-reaching consequences for the neurohumoral transmission theory.

DALE had not yet formed a clear notion of neurohumoral transmission at that time. In a retrospective lecture he characterized the position around 1914 as follows: "Two substances were known with actions suggestively reproducing those of the two main branches of the autonomic nervous system: both were very unstable in the body and their actions were consequently of a fleeting character and one of them was already known to occur as a natural hormone. These properties would fit them to act as mediators of autonomic impulses to effector cells if there were any acceptable evidence of their liberation at the nerve endings. The actors were named and the parts allotted. A preliminary hint of a plot had indeed been given ten years earlier but had been completely forgotten (ELLIOTT's hint that adrenaline might be the sympathetic transmitter), but only direct and unequivocal evidence could ring up the curtain and this was not to come until 1921."

In 1921 OTTO LOEWI began his series of experiments, of classical simplicity and economy of means, which established the neurohumoral transmission concept. The subsequent work by DALE and his colleagues was complementary to that of LOEWI. Whereas LOEWI had generated a brilliant general idea, DALE gave it life and substance by applying it to autonomic and somatic nerves and building up an integrated system of neurohumoral transmission in which the concepts of muscarine and nicotine actions of acetylcholine played an important role. Their main discoveries, made in rapid succession, can be summarised as follows. Acetylcholine is the chemical transmitter of vagus effects to the stomach (DALE and FELDBERG, 1934). Acetylcholine mediates transmission of the nerve impulse from sympathetic postganglionic nerve endings to sweat glands (DALE and FELDBERG, 1934). (In this context DALE suggested the words "cholinergic" and "adrenergic" to provide a functional classification of nerve endings). Acetylcholine is involved in the transmission of the impulse from preganglionic nerve endings of the splanchnic to the adrenal medulla (FELDBERG and colleagues, 1931–1934). Acetylcholine is the transmitter of the preganglionic stimulus in the superior cervical ganglion (FELDBERG and GADDUM, 1934). Acetylcholine is the transmitter of the motor impulse to striated muscle (DALE, FELDBERG and VOGT; BROWN, 1936). Although the idea that the short-latency nicotinic transmissions at ganglia and the motor end plate could be humorally mediated was resisted at first, especially by electrophysiologists, the neurohumoral theory became widely accepted by about 1937.

The work of LOEWI and DALE led to a transformation of physiological and pharmacological thinking. The idea, which had persisted since CLAUDE BERNARD, that curare blocked motor nerve endings, and the related ideas that adrenaline and pilocarpine stimulated sympathetic and parasympathetic nerve endings and that atropine blocked the latter, had to be abandoned. Equally important were the consequences for clinical medicine. DALE seemed almost surprised at the practical applications of his experimental work, especially as he was brought up on EHRLICH's ideas that only drugs like the chemotherapeutic agents which dealt with the "causes" of disease rather than with "symptoms" were of real value. It now became possible to evolve drugs by rational processes which interfered with neurohumoral transmission and in this way could save or prolong life, without being truly curative. Examples of drugs developed as direct or indirect extensions of DALE's work are the anticholinesterases in myasthenia gravis, tubocurarine as muscle relaxant in anaesthesia and ganglion- and adrenergic neurone blocking-drugs in the treatment of hypertension. It is of historical interest that one of the first demonstrations of ganglion block was given in a paper published by BURN and DALE in 1915 in which they showed that tetraethylammonium antagonised the ganglion stimulant effect of tetramethylammonium. Some thirty years later tetraethylammonium was used clinically as the first effective ganglion blocker for the treatment of hypertension.

Oxytocic Drugs

DALE made fundamental contributions to this field in which he remained interested throughout his career. He discovered the oxytocic effect of posterior pituitary extract in 1906. He also discovered independently (but simultaneously with KEHRER and CUSHNY) the dual effect of adrenaline on cat uterus, depending on hormonal status. He was inclined at first to attribute the oxytocic and pressor effects of posterior pituitary extract to a single active substance (applying the logical principle of *Occam's razor*), but when DUDLEY prepared separate oxytocic and pressor fractions he concluded that two separate substances were involved. This led to a controversy with ABEL, who maintained that the two active substances were artifacts, and that the natural hormone consisted of a single macromolecule to which two different prosthetic groups were attached. ABEL's idea later proved to be basically correct, but DALE pointed out that the essential feature was whether the two hormones were always secreted together into the blood stream, which has been shown experimentally not to be the case.

The further development of the ergot story was also greatly influenced by DALE. Ergotoxine had been isolated from ergot extract and its presence could be assessed by the adrenaline reversal reaction, but it soon became clear that obstetricians did not trust this pharmacological test. They persisted in regarding samples of liquid extract of ergot as active even when pronounced pharmacologically inactive. The discrepancy became resolved, under DALE's guidance, when CHASSAR MOIR in Oxford demonstrated by means of intra-uterine balloons in parturient women that ergotoxine-free extracts nevertheless retained uterus-contracting activity. This led to the isolation by DUDLEY of ergometrine, which has strong uterotonic but little antiadrenaline activity. Ergometrine has since remained the standard drug employed in postpartum haemorrhage.

Biological Standardisation

DALE made significant contributions to many different fields, as can be seen from his list of publications. He has been called "perhaps the last of the polymaths of medical research" (F. G. Young). He exerted a powerful influence on his surroundings and some of his indirect contributions through others are as significant as his own research. An example of Dale interacting with his surroundings is provided by his work on biological standardisation, of which he says: "for many years it made a claim on time and thought which I would willingly have given to more fundamental but not, I think, more important research activities". Their importance was threefold. First, by insisting on the essentially comparative nature of biological assay and the necessity for comparison with a stable standard preparation (a principle first recognized by Ehrlich in relation to diphtheria antitoxin), he made sure that reliable samples of insulin, posterior pituitary and digitalis became available to the medical profession. Before the introduction of international standards, commercial preparations of posterior pituitary varied by as much as 80-fold in activity. Second, Dale gave the clearest possible formulation of the principles underlying biological standardization. Third, and perhaps most important, under his direct influence a school of bioassay arose (Trevan, Gaddum, Burn, etc.) which had a predominant effect on the introduction of statistical measuring techniques into pharmacology and medicine. Much of today's emphasis on the quantitative assessment of drug action can be traced back to the influence of this group.

Dale's Scientific Philosophy

Dale's philosophy of science was grounded in the optimistic traditions of the nineteenth century. He saw no contradiction between science as a disinterested search for truth and science as a social activity, and felt confident that the pursuit of pure research would ultimately lead to the betterment of the condition of human life. Temperamentally he did not believe in the organized direction of research, holding that freedom and opportunity, rather than organization, provided the conditions for the highest types of research and thus in the end for the greatest services to mankind. On the other hand he points out that the government support of research in Great Britain as administered by the three Research Councils has been beneficial and has not endangered the freedom of science.

Underneath this general air of optimism he was, however, deeply troubled by some of the problems of our time and their impact upon science, and he expressed his views on them forcefully and courageously. Before the war, he took a strong stand on the persecution of scientists and gave unqualified help and support to refugees coming to England. After the war, he was equally eager to re-open international communication, being a great believer in the supranational role of science.

He was perturbed by the moral problems which scientists working on secret war research may have to face, and at one stage raised the question whether scientists of all countries could not bind themselves by something analogous to the Hippocratic Oath. He took a particularly strong stand on the question of secrecy. The following extract is from his Presidential Address to the Royal Society, 1945.

"I think that we, as scientists, should make it clear to the world that, if national military secrecy were allowed thus progressively to encroach upon the freedom of science, even if civilization should yet for a while escape the danger of final destruction, a terrible, possibly mortal wound would have been inflicted on the free spirit of science itself, to the immeasurable loss of what it stands ready to offer to a wider world."

Biographical Note

HENRY HALLETT DALE was born in 1875 in London, the son of a business-man. He lived in London most of his life, except for a period of study at Cambridge and for his last years, when he and his wife lived in a nursing home in Cambridge close to their daughter's family. He married his cousin ELLEN HALLETT in 1904; they were a devoted couple until her death in 1967. Their only son, a doctor, died young in Canada. Of their two daughters, one married a doctor, the other a chemist, ALEC TODD, later Lord TODD, who, like his father-in-law, was to receive a Nobel Prize.

DALE was a scholar and research student at Trinity College Cambridge from 1894–1900. He was especially attracted to the department of physiology which at that time contained MICHAEL FOSTER, GASKELL and LANGLEY, and considered GASKELL's advanced lecture course to have been the most inter-esting he ever attended. He graduated, with first class honours in physiology, in 1898 and subsequently obtained a studentship at Trinity which had become vacant through the departure of the physicist ERNEST RUTHERFORD. After two years' work with LANGLEY, DALE moved to St. Bartholomew's Hospital, London, to complete his medical training. After graduation he had the choice of a medical post as house physician or a research post with Starling at University College London, and chose the latter.

During this period, DALE worked for a time with EHRLICH in Frankfurt, but apparently experienced some difficulty in communicating with him (DALE used to relate how he and OTTO LOEWI paid a joint visit to EHRLICH's office; after leaving, having listened to EHRLICH's explanations of his work, LOEWI said to DALE: haben Sie ein Wort verstanden? DALE: nein; LOEWI: ich auch nicht). Yet DALE's appreciation of EHRLICH grew steadily with the years, and he eventually edited a complete edition of EHRLICH's scientific writings.

In 1904, at the age of 29, DALE, finding academic prospects bleak and wanting to get married, took up a research post with the pharmaceutical firm of Burroughs Wellcome. Although this was a rather unusual step for that time, he had no cause to regret the decision of taking up an industrial appointment and spent ten of his most productive years with Wellcome. He never again entered university service but in 1914, the year he became a Fellow of the Royal Society, took up a civil service post with the newly formed Medical Research Committee, later Medical Research Council, and eventually became the director of the Council's laboratories at Hampstead.

His personal laboratory there, affectionately known as F4, became world-famous and attracted many distinguished scientists. The laboratory consisted essentially of one largish room in which two or three experiments could be carried out simultaneously. Any experiment planned for next day had to be announced to eagle-eyed COLLISON, who saw to it that everything was perfectly prepared and functional. In these restricted surroundings, with a few permanent collaborators and one or two visitors from abroad, were carried out some of the most distinguished physiological and pharmacological researches of the century.

In old age DALE was considered the outstanding scientist of his country. He had achieved the highest scientific honours, including that of president of the Royal Society in the crucial war years, and had received the Nobel Prize, the Order of Merit and countless other distinctions. Yet he remained personally completely simple, approachable, interested and kind until his death on 23 July 1968 at the age of 93.

References

1900. DALE, H. H.: On some numerical comparisons of the centripetal and centrifugal medullated nerve-fibres arising in the spinal ganglia of the mammal. J. Physiol. (Lond.) **25**, 196.

1901. — Galvanotaxis and chemotaxis of ciliate infusoria. J. Physiol. (Lond.) **26**, 291.

— Observations, chiefly by the degeneration method, on possible efferent fibres in the dorsal nerve-roots of the toad and frog. J. Physiol. (Lond.) **27**, 250.

1904. — The Islets of Langerhans of the pancreas. (Abstr.) Proc. roy. Soc. **73**, 84. On the "Islets of Langerhans" in the pancreas. Phil. Trans. B **197**, 25.

1905. BAINBRIDGE, F. A., DALE, H. H.: The contractile mechanism of the gall-bladder and its extrinsic nervous control. J. Physiol. (Lond.) **33**, 138.

DALE, H. H.: The physiological action of chrysotoxin. (Prel. Com.) J. Physiol. (Lond.) **32**, 58P.

1906. BARGER, G., DALE, H. H.: Die Mutterkornalkaloide. Arch. Pharm. (Weinheim) **244**, 550.

DALE, H. H.: On some physiological actions of ergot. J. Physiol. (Lond.) **34**, 163.

1907. BARGER, G., DALE, H. H.: Ergotoxine and some other constituents of ergot. Biochem. J. **2**, 240.

1909. — — Über Mutterkorn. Naunyn-Schmiedebergs Arch. exp. Path. Pharmak. **61**, 113.

— — The water soluble active principles of ergot. J. Physiol. (Lond.) **38**, 77P.

DALE, H. H., LAIDLAW, P. P., SYMONS, C. T.: Acceleration of the mammalian heart-beat by stimulation of the vagus nerve. J. Physiol. (Lond.) **39**, 13 P.

— — The action of an active principle from Apocynum. Heart **1**, 138.

— — The action of some diuretics. Proc. roy. Soc. Med., therap. Sect. **3**, 38.

— — The physiological action of primary fatty amines. J. Physiol. (Lond.) **38**, 22 P.

DALE, H. H.: Note on nutmeg poisoning. Proc. roy. Soc. Med., therap. Sect. **2**, 69.

— DIXON, W. E.: The action of pressor amines produced by putrefaction. J. Physiol. (Lond.) **39**, 25.

— The action of extracts of the pituitary body. Biochem. J. **4**, 427.

1910. BARGER, G., DALE, H. H.: A third active principle in ergot extracts. (Prel. note) Proc. chem. Soc. (Lond.) **26**, 218.

— — The presence in ergot and physiological activity of β-imidazolylethylamine. (Prel. Com.) J. Physiol. (Lond.) **40**, 38 P.

DALE, H. H., LAIDLAW, P. P., SYMONS, C. T.: A reversed action of the vagus on the mammalian heart. J. Physiol. (Lond.) **41**, 1.

— — The physiological aticon of β-iminazolylethylamine. J. Physiol. (Lond.) **41**, 318

BARGER, G., DALE, H. H.: Die physiologische Wirkung einer Secalebase und deren Identifizierung als Imidazolyläthylamin. Zbl. Physiol. **24**, 885.

— — 4-β-aminoethylglyoxaline (β-iminazolylethylamine) and the other active principles of ergot. J. chem. Soc. Trans. **97**, II, 2592.

DALE, H. H.: The active principles of ergot. Brit. med. J. II, 1610.

— The active principles of ergot. Proc. Internat. Physiol. Congr., Vienna, Sept., 27–30.

1911. BARGER, G., DALE, H. H.: Chemical structure and sympathomimetic action of amines. J. Physiol. (Lond.) **41**, 19.

DALE, H. H., LAIDLAW, P. P.: Further observations on the action of β-iminazolylethylamine. J. Physiol. (Lond.) **43**, 182.

— — Note on a reversed action of the chorda tympani on salivary secretion. J. Physiol. (Lond.) **43**, 196.

BARGER, G., DALE, H. H.: β-iminazolylethylamine a depressor constituent of intestinal mucosa. J. Physiol. (Lond.) **41**, 499.

1912. DALE, H. H., LAIDLAW, P. P.: A method of standardising pituitary (infundibular) extracts. J. Pharmacol. **4**, 75.

— — A simple coagulometer. J. Path. Bac. **16**, 351.

— — The physiological action of cytisine, the active alkaloid of laburnum (Cytisus laburnum). J. Pharmacol. **3**, 205.

— — A method of preparing secretin. J. Physiol. (Lond.) **44**, 11 P.

— — Some actions of pilocarpine and nicotine. (Prel. Com.) J. Physiol. (Lond.) **44**, 12 P.

— — The significance of the suprarenal capsules in the action of certain alkaloids. J. Physiol. (Lond.) **45**, 1.

— The anaphylactic reaction of plain muscle in the guinea-pig. (Prel. Com.) J. Physiol. **45**, 27 P. Same title. J. Pharmacol. **4**, 167.

— Ergot and its active principles. J. Amer. pharm. Ass., Nov.

1913. CARR, F. H., DALE, H. H.: Ergot and its preparations; a critical review of the requirements of the British Pharmacopoeia. Pharm. J. 4. s., **37**, 130.

— — Ergot, and its preparations. Med. Pr, N.S. **96**, 254.

DALE, H. H.: On the action of ergotoxine; with special reference to the existence of sympathetic vasodilators. J. Physiol. (Lond.) **46**, 291.

— The effect of small variations in concentration of Ringer's solution on the response of isolated plain muscle. (Prel. Com.) J. Physiol. (Lond.) **46**, 19 P.

— The effect of varying tonicity on the anaphylactic and other reactions of plain muscle. J. Pharmacol. **4**, 517.

1914. — EWINS, A. J.: Choline-esters and muscarine. (Prel. Com.) J. Physiol. (Lond.) **48**, 24 P.

BARGER, G., DALE, H. H.: Liver nitrogen in anaphylaxis. Biochem. J. **8**, 670.

Dale, H. H.: The occurrence in ergot and action of acetylcholine. (Prel. Com.) J. Physiol. (Lond.) **48**, 3P.

— The action of certain esters and ethers of choline, and their relation to muscarine. J. Pharmacol. **6**, 147.

1915. Barger, G., Dale, H. H.: Note on a supposed soluble toxin, produced in artificial culture by the bacillus of malignant oedema. Brit. med. J. II, 808.

Dale, H. H.: A preliminary note on chronic poisoning by emetine. Brit. med. J. II, 895.

Burn, J. H., Dale, H. H.: The action of certain quarternary ammonium bases. J. Pharmacol. **6**, 417.

Dale, H. H.: The physiology of the thyroid gland. Practitioner **94**, 16.

1916. — Walpole, G. S.: Some experiments on factors concerned in the formation of thrombin. Biochem. J. **10**, 331.

— Hartley, P.: Anaphylaxis to the separated proteins of horse-serum. Biochem. J. **10**, 408.

— Treatment of carriers of amoebic dysentery. Note on the use of the double iodide of emetine and bismuth. Lancet II, 183.

1917. — Dobell, C.: Experiments on the therapeutics of amoebic dysentery. J. Pharmacol. **10**, 399.

Barger, G., Dale, H. H., Durham, F. M.: "Collosol" cocaine. Lancet II, 825.

Dale, H. H.: The treatment of amoebic dysentery. (Letter) Lancet I, 780.

1918. — Richards, A. N.: The vasodilator action of histamine and of some other substances. J. Physiol. (Lond.) **52**, 110.

1919. — Contribution to general discussion on shock (introduction by W. M. Bayliss). Proc. roy. Soc. Med. **12**, 4.

— Laidlaw, P. P.: A simple method of short-circuiting the portal circulation. J. Physiol. (Lond.) **52**, 351.

— — Histamine shock. J. Physiol. (Lond.) **52**, 355.

— Royal Society Croonian Lecture. The biological significance of anaphylaxis. (Abstract) Proc. roy. Soc. B **90**, 556.

— Laidlaw, P. P., Richards, A. N.: The action of histamine: its bearing on traumatic toxaemia as a factor in shock. In: Med. Res. Com. Special Report, Ser. No. 26, 8.

Dakin, H. D., Dale, H. H.: Chemical structure and antigenic specificity. A comparison of the crystalline egg-albumins of the hen and the duck. Biochem. J. **13**, 248.

Richards, A. N., Dale, H. H.: Experiments on the action of histamine. Proc. path. Soc. Philad., N.S. **22**, 32.

Dale, H. H.: Supplementary note on histamine shock. Med. Res. Com. Spec. Rep. Ser. No. 26, 15.

1920. — Royal Society Croonian Lecture. The biological significance of anaphylaxis. Proc. roy. Soc. B **91**, 126.

— Evans, C. L.: Colorimetric determination of the reaction of blood by dialysis. J. Physiol. (Lond.) **54**, 167.

— The use of colloidal preparations in medicine. J. Soc. chem. Ind. (Lond.) **39**, 211.

— Conditions which are conducive to the production of shock by histamine. Brit. J. exp. Path. **1**, 103.

— The Herter Lectures. Baltimore 1919. 1. Capillary poisons and shock. Johns Hopk. Hosp. Bull. **31**, 257. 2. Anaphylaxis. Johns Hopk. Hosp. Bull. **31**, 310. 3. Chemical structure and physiological action. Johns Hopk. Hosp. Bull. **31**, 373.

— Contribution to: Discussion on the value of alcohol as a therapeutic agent. Proc. roy. Soc. Med., therap. Sect. **13**, 31.

1921. — Recent tendencies in chemotherapy. President's Address. Proc. roy. Soc. Med., therap. Sect. **14**, 7.

— Dudley, H. W.: The physiological action of N-methylhistamine and of tetrahydropyrido-3.4-iminazole ("imidazolisopiperidin" of Fränkel). J. Pharmacol. **18**, 103.

— — On the pituitary active principles and histamine. J. Pharmacol. **18**, 27.

1921. — Kellaway, C. H.: Anaphylaxis and immunity in vitro. (Prel. Com.) J. Physiol. (Lond.) **54**, 142P.

— Anaphylaxis. St. Bart. Hosp. J. **28**, 163.

— Note on the reversal of vagus action by quinidine, as seen in the heart of the cat. Heart **9**, 87.

— The nature and cause of wound shock. Harvey Society Lectures, New York, p. 27.

— The use of alcohol in medicine. Brit. J. Inebr. **19**, 27.

— Hill, L.: Anaesthesia with nitrous oxide and oxygen under pressure. Lancet II, 326.

— The work of the National Institute for Medical Research. Lancet II, 112 (and The Fight Against Disease (Res. Def. Soc.) October 1921).

— Anaphylatoxin. (Abstract) Brit. med. J. II, 689.

1922. — Francis Arthur Bainbridge, 1874–1921. (Obit. Not.) Proc. roy. Soc. B **93**, 24.

— White, C. F., et al.: Report on an experimental and clinical comparison of the therapeutic properties of different preparations of 914 (neosalvarsan). Lancet I, 779. (Based on experiments made in conjunction with J. H. Burn, Florence M. Durham, Juliette Marchal and C. H. Mills.)

— Evans, C. L.: Effects on the circulation of changes in the carbondioxide content of the blood. J. Physiol. (Lond.) **56**, 125.

— Spiro, K.: Die wirksamen Alkaloide des Mutterkorns. Naunyn-Schmiedebergs Arch. exp. Path. Pharmak. **95**, 337.

— Kellaway, C. H.: Anaphylaxis and anaphylatoxins. Phil. Trans. B. **211**, 273.

Burn, J. H., Dale, H. H.: Reports on biological standards. 1. Pituitary extracts. Spec. Rep. Ser. med. Res. Coun. No 69. (Lond.)

Dale, H. H.: Britisch Medical Association Lecture. Specific sensitiveness and anaphylaxis. Brit. med. J. I, **45**.

1923. — The value of ergot in obstetrical and gynaecological practice; with special reference to its present position in the British Pharmacopoeia. Proc. roy. Soc. Med., therap. Sect. **16**, 1.

— Colloidal preparations. (Letter) Brit. med. J. I, 786.

— Hadfield, C. F., King, H.: The anaesthetic action of pure ether. Lancet I, 424.

— The Oliver-Sharpey Lectures to the Royal College of Physicians. The activity of the capillary blood vessels, and its relation to certain forms of toxaemia. Brit. med. J. I, 959 and 1006.

— Chemotherapy. Physiol. Rev. **3**, 359.

— The physiology of insulin. Lancet I, 989.

1924. — The search for specific remedies. Proc. roy. Instn. G. B. **23**, 635.

Burn, J. H., Dale, H. H.: On the location and nature of the action of insulin. J. Physiol. (Lond.) **59**, 164.

Dale, H. H.: Memo. on the present position with regard to the testing of salvarsan etc. League of Nations. June, 1925.

— Presidential Address to Section of Physiology, Brit. Assoc., Toronto, 1924. Progress and prospects in chemotherapy. Brit. med. J. II, 219–223.

1925. — The circulation of blood in the capillary vessels. Proc. roy. Instn. B.B. **24**, 623.

1926. — Drummond, J. C., Henderson, L. J., Hill, A. V.: Lectures on certain aspects of biochemistry, p. 313. London: Univ. London. Press.

Burn, J. H., Dale, H. H.: The vaso-dilator action of histamine and its physiological significance. J. Physiol. (Lond.) **61**, 185.

Dale, H. H., Gasser, H. S.: The pharmacology of denervated mammalian muscle. I. The nature of the substances producing contracture. J. Pharmacol. **29**, 53.

Gasser, H. S., Dale, H. H.: The pharmacology of denervated mammalian muscle. II. Some phenomena of antagonism and the formation of lactic acid in chemical contracture. J. Pharmacol. **28**, 287.

Best, G. H., Dale, H. H., Hoet, J. P., Marks, H. P.: Oxidation and storage of glucose under the action of insulin. Proc. roy. Soc. B **100**, 55.

Dale, H. H.: Introduction. The biological standardisation of insulin, p. 398. Geneva: League of Nations Health Organisation.
— Arthur Robertson Cushny. (Obit. Not.) Arch. int. Pharmacodyn. 32, 1.
— Arthur Robertson Cushny. (Obit. Not.) Proc. roy. Soc. B 100, 19.
— The scientific and industrial problems presented by the hormones—the natural drugs of the body. The experimental study and use of hormones. J. Soc. chem. Ind. (Lond.) 45, 235.
1927. — Dudley, H. W.: An active constituent of the preparation called Glukhorment. Brit. med. J. II, 1027.
— Richards, A. N.: The depressor (vaso-dilator) action of adrenaline. J. Physiol. (Lond.) 63, 201.
Best, C. H., Dale, H. H., Dudley, H. W., Thorpe, W. V.: The nature of the vaso-dilator constituents of certain tissue extracts. J. Physiol. (Lond.) 62, 397.
Dale, H. H.: Die Theorie der Insulinwirkung. Karlsbader ärztliche Vorträge 9, 393.
1928. — Schuster, E. H. J.: A double perfusion pump. J. Physiol. (Lond.) 64, 356.
— Discussion on the action of synthalin. Proc. roy. Soc. Med., therap. Sect. 21, 527.
— Some reactions of pharmacology on pharmacy. Pharm. J. 120, 245.
1929. — Croonian Lectures to the Royal College of Physicians: Some chemical factors in the control of the circulation. 1. Introduction, vaso-motor hormones. Lancet I, 1179. 2. Local vasodilator reactions—histamine. Lancet I, 1233. 3. Local vasodilator reactions. Histamine cont., acetylcholine, conclusion. Lancet I, 1285.
— Dudley, H. W.: The presence of histamine and acetylcholine in the spleen of the ox and the horse. J. Physiol. (Lond.) 68, 97.
1930. Corkill, A. B., Dale, H. H., Marks, H. P.: The respiratory quotient of the eviscerated spinal cat. J. Physiol. (Lond.) 70, 86.
Dale, H. H., Gaddum, J. H.: Reactions of denervated voluntary muscle, and their bearing on the mode of action of parasympathetic and related nerves. J. Physiol. (Lond.) 70, 109.
— In Memoriam Rudolf Magnus 1873–1927. Introductory note to Lane Lectures on Experimental Pharmacology and Medicine, by Rudolf Magnus. Stanford: University Press.
1931. — Dudley, H. W.: Enthält das normale Blut Acetylcholin? Mit Berücksichtigung der Arbeit von Kapfhammer und Bischoff. Hoppe-Seylers. Z. physiol. Chem. 198, 85.
— The effect of research on curative medicine. Stephen Paget Memorial Lecture. Brit. med. J. I, 1076.
— Presidential Address to Section of Physiology, Brit. Assoc. (Centen. Meeting), London. (Introducing discussion on biological nature of viruses.) Ann. Rep. Brit. Assn., p. 172.
— British Science Guild. The Norman Lockyer Lecture 1931. Biology and civilization. Nature (Lond.) 128, 897 (1931).
1932. — Über Kreislaufwirkungen körpereigener Stoffe. (Deut. Kongr. inn. Med., Wiesbaden, 1932). Naunyn-Schmiedebergs Arch. exp. Path. Pharmak. 167, 21.
Bauer, W., Dale, H. H., Poulsson, L. T., Richards, D. W.: The control of circulation through the liver. J. Physiol. (Lond.) 74, 343.
Dale, H. H., Marble, A., Marks, H. P.: The effects on dogs of large doses of calciferol (Vitamin D). Proc. roy. Soc. B 111, 522.
— Notes on communication by J. Chassar Moir on: The action of ergot preparations on the puerperal uterus. Brit. med. J. I, 1119.
— Presidential Address (Section of Physiology, Brit. Med. Ass., Centenary Meeting). The relation of physiology to medicine in research and education. Brit. med. J. II, 1043.
— Some therapeutic problems of the future. (Harrison Memorial Lecture to the Pharmaceutical Society) Pharm. J. 129, 515.
1933. — Virus diseases (Substance of a lecture entitled: Ultramicroscopic organisms and the troubles which they cause, delivered at Bedford College, London, on 1 March 1933). Nature (Lond.) 131, 370.

1933. — The Dohme Lectures (3). Baltimore, 1933. Progress in autopharmacology. A survey of present knowledge of the chemical regulation of certain functions by natural constituents of the tissues. Johns Hopk. Hosp. Bull. **53**, 297.

— Academic and industrial research in the field of therapeutics. An address delivered at the opening ceremony of Merck Research Laboratory, Rahway, N.J., 25 April. Science **77**, 521.

1934. — FELDBERG, W.: The chemical transmitter of effects of the gastric vagus. J. Physiol. (Lond.) **80**, 16P.

— Nomenclature of fibres in the autonomic system and their effects. J. Physiol. (Lond.) **80**, 10P.

— The Linacre Lecture, Cambridge, 1934. Chemical transmission of the effects of nerve impulses. Brit. med. J. I, 835.

— FELDBERG, W.: The chemical transmitter of vagus effects to the stomach. J. Physiol. (Lond.) **81**, 320.

— — The chemical transmitter of nervous stimuli to the sweat glands of the cat. J. Physiol. (Lond.) **81**, 40P.

— — Chemical transmission at motor nerve endings in voluntary muscle. J. Physiol. (Lond.) **81**, 39P.

— — The chemical transmission of secretory impulses to the sweat glands of the cat. J. Physiol. (Lond.) **82**, 121.

— Chemical ideas in medicine and biology. Science **80**, 343. Address given at the opening ceremony of the Eli Lilly Research Laboratories, 11 October 1934.

1935. — Walter Ernest Dixon Memorial Lecture, 1934. Pharmacology and nerve-endings. Proc. roy. Soc. Med., therap. Sect. 1935, **28**, 319.

— Note (on nomenclature) appended to FELDBERG, W., and GUIMARAIS, J. A. J. Physiol. (Lond.) **83**, 43P.

— Contribution to discussion on traumatic shock. Proc. roy. Soc. Med., surg. Sect. **28**, 1493.

— Viruses and heterogenesis. An old problem in a new form. Huxley Memorial Lecture, 2 May 1935. London: Macmillan & Co.

— Discussion. The new ergot alkaloid. Science **82**, 99.

BROWN, G. L., DALE, H. H.: The pharmacology of ergometrine. Proc. roy. Soc. B **118**, 446.

DALE, H. H.: Harold Ward Dudley. 1887–1935. Obit. Not. Roy. Soc. 1935, **1**, 595.

— Die Pharmakologie des Mutterkorns. Schweiz. med. Wschr. **65**, 885. With Discussion by Profs. A. STOLL, E. ROTHLIN, W. LOEFFLER and Sir HENRY DALE. Schweiz. med. Wschr. **65**, 1077.

— The Harveian Oration before the Royal College of Physicians. Some epochs in medical research. London: H. K. Lewis & Co. 1935. Also Brit. med. J. **1935**, II, 771.

1936. — Recent developments in the pharmacology of ergot, especially of ergometrine. Lecture Ned. Vereeniging voor Physiologie en Pharmacologie 5. III. 1936. Acta brev. neerl. Physiol. **6**, 1.

BROWN, G. L., DALE, H. H.: Perfusion with solution of haemolysed red corpuscles. J. Physiol. (Lond.) **86**, 42P.

DALE, H. H., FELDBERG, W., VOGT, M.: Release of acetylcholine at voluntary motor nerve endings. J. Physiol. (Lond.) **86**, 353.

BROWN, G. L., DALE, H. H., FELDBERG, W.: Chemical transmission of excitation from motor nerve to voluntary muscle. J. Physiol. (Lond.) **87**, 42P.

— — — Reactions of the normal mammalian muscle to acetylcholine and to eserine. J. Physiol. (Lond.) **87**, 394.

DALE, H. H.: Some recent extensions of chemical transmission. Cold Spr. Harb. Symp. quant. Biol. **4**, 143.

GADDUM, J. H. (Eingeleitet von DALE, H. H.): Gefäßerweiternde Stoffe der Gewebe. Leipzig: Georg Thieme.

1937. Dale, H. H.: Transmission of nervous effects by acetylcholine. Harvey Lectures, New York, 1936–1937. Lecture delivered 20 May 1937. Harvey Lect. 1937, **32**, 229.
— International standard progesterone. Bull. Hth Org. L. o.N. **6**, 892.
— Some recent extensions of the chemical transmission of the effects of nerve impulses. Nobel Lecture, December 1936. Les Prix Nobel en 1936 (Stockholm, 1937).
— Medicine as an experimental science. Address at the celebration of the 150th Anniversary of the Founding of the College of Physicians of Philadelphia, 14 May 1937. Trans. Coll. Phycns Philad. 1937, **4**, (Suppl.) 4th. Ser. p. 52.
— Du Bois-Reymond and chemical transmission. J. Physiol. (Lond.) **91**, 4P.
1938. — The William Henry Welch Lectures, 1937: Acetylcholine as a chemical transmitter of the effects of nerve impulses. 1. History of ideas and evidence. Peripheral autonomic actions. Functional nomenclature of nerve fibres. 2. Chemical transmission at ganglionic synapses and voluntary motor nerve endings. Some general considerations. J. Mt Sinai Hosp. 1938, **4**, 401.
— Acetylcholine as transmitter of the effects of nerve impulses. I. Pavlov Number, Physiol. J., U.S.S.R. **24**, 116.
— Natural chemical stimulators. Edinb. med. J., N.S. (IVth) **45**, 461. (The Sharpey Schafer Memorial Lecture delivered to the Faculty of Medicine, University of Edinburgh, 1937).
— Chemical agents transmitting nervous excitations. Irish J. med. Sci., June 1938, 1 (John Mallet Purser Lecture, 1938).
— The future of medicine. Report of an address given at the opening of the new session, Guy's Hospital, on 14 October 1938. Guy's Hosp. Gaz. **52**, 396.
— Discussion on the scope and future of teaching and research in pharmacology. Introduction to a Discussion Meeting of Pharmacologists. Proc. XVIth International Physiological Congress, Zurich.
1939. — John Jacob Abel. 1857–1938. Obit. Not. Roy. Soc. 1939, **2**, 577.
— Biological standardisation. (Address given to the Annual General Meeting of the Society of Public Analysts, 1939) Analyst **64**, 554.
— Chemical mediation in the peripheral nervous system and its relation to endocrine organs. Comptes Rendus. IIIe Congrès Neurologique International, Copenhagen, 1939, p. 37. Also (in French) Rev. neurol. 1940, **72**, 347.
— Physiology of the nervous system. Science **90**, 393. Discussion.
— Address to the Assembly of the Faculties of University College, London, 6 July 1939. Annual Report, University of London, University College, 1939–1940, p. 46.
1940. — George Barger. Obit. Not. Roy. Soc., 1940, **4**, 63. Lancet **1939**, I 116. Also Nature (Lond.) 1939, **143**, 107.
— Sir Frederick Andrewes. Dictionary of National Biography. Suppl. 1931–1940, p. 14.
1941. — Sir Frederick Banting. Obit. Not. Brit. med. J. I, 383.
— Patrick Playfair Laidlaw. Obit. Not. Roy. Soc. 1941, **3**, 427.
— A new era in medicinal treatment. Mem. Proc. Manchr. Lit. Phil. Soc. 1941, **84**, 42.
— Address of the President, Anniversary Meeting, 1941. Proc. roy. Soc. B **130**, 227.
1942. — The mode of action of chemotherapeutic agents. A discussion held by the Biochemical Society, 29 November 1941. Biochem. J. **36**, 3.
— Wartime arrangements for international biological standards and the new standard for pituitary (posterior lobe) preparations. Brit. med. J. II, 385.
— Edgar Vincent, Viscount D'Abernon, 1857–1941. Obit. Not. Roy. Soc. **4**, 83.
1943. — A prospect in therapeutics. (Frederick Price Lecture to the Royal College of Physicians, Edinburgh, 2 July 1943) Brit. med. J. II, 411.
— Address of the President, Anniversary Meeting, 1942. Introduction to Addresses at Isaac Newton Tercentenary. Proc. roy. Soc. A **181**, 211 and 224.
1944. — Address of the President, 1943. Proc. roy. Soc. A 1944, **182**, 217; Proc. roy. Soc. B **132**, 1.
— Henry George Lyons, 1864–1944. Obit. Not. Roy. Soc. 1944, **4**, 795.

1944. — International activities in science. Nature (Lond.) **154**, 725.

— The natural history and chemistry of drugs. Address to The Pharmaceutical Society of Great Britain, 13 January 1944. Pharm. J. **152**, 34.

— Discussion (with J. B. Buxton and G. W. Pickering) on principles and relationships involved in medical and veterinary education. Proc. roy. Soc. Med. **38**, 119.

1945. — Address of the President, 1944. Proc. Roy. Soc. B 1945, **132**, 333.

— A hitherto unpublished letter of Isaac Newton. Nature (Lond.) **156**, 193.

— The mission of science. Nature (Lond.) **156**, 677.

1946. — Address of the President, 1945. Proc. roy. Soc. B 1946, **133**, 123.

— Experiment in medicine. History of Science Lecture, Cambridge, 2, March 1946. Camb. Hist. J. **8**, 166.

— Electric fishes. (Abstract) Discourse at Royal Institution. Proc. Roy. Instn. G.B. **33**, 662.

— Diabetic Association. Silver Jubilee Celebration of the Discovery of Insulin. Memorial Address. Diabetic J. **4**, 392.

1947. — Science in war and peace. The Presidential Address to the British Association Meeting, Dundee, 27 August 1947. The Advanc. Sci. **4**, 151. Also Nature (Lond.) **160**, 283.

— Science and medicine. (Broadcast) Chapter in Atomic Challenge, B.B.C. Symposium p. 107. London: Winchester Publications.

— The Pilgrim Trust Lecture, 1946. The freedom of science. Proc. Amer. Phil. Soc. 1947, **91**, 64.

— Walter Bradford Cannon. Obit. Not. Roy. Soc., 1947, **5**, 407.

— The part of chemistry in the new therapeutics (XI International Congress of Pure and Applied Chemistry, 1947, London, 17 to 24 July) Suppl. to Chemistry and Industry, 1948.

1948. — Chemistry and medicinal treatment. Third Dalton Lecture, 1947. Royal Institute of Chemistry, 1948.

— Transmission of effects from the endings of nerve fibres. Nature (Lond.) **162**, 558.

— Antihistamine substances. (Opening to a discussion in the Section of Pharmacology, B.M.Ass., 1948) Brit. med. J. II, 281.

— Accident and opportunism in medical research. (Popular Lecture at the Annual Meeting, B.M.Ass., 1948) Brit. med. J. II, 451.

— The physiological basis of neuromuscular disorders. (Opening a discussion in the Section of Physiology, B.M.Ass., 1948) Brit. med. J. II, 889.

— Frederick Gowland Hopkins. Obit. Not Roy. Soc. 1948, **6**, 115.

— Medicinal treatment—then and now. B.B.C. Broadcast, 4 September 1946. Science Survey, 1948, p. 239. London: Sampson Low, Marston & Co. Ltd.

— Atomic energy. Its international implications. Introductory Chapter The Objective, p. 15. Royal Institute of International Affairs, London. Oxford: University Press.

— Research and organisation. (Conference at Manchester on the place of Universities in the Community, 10 May 1947). The Advanc. Sci. 1948, **5**, 11.

1949. — Contribution to Discussion. Annual Conference of Atomic Scientists' Association held at Beaver Hall, London, on 30 October 1948. Atomic Scientists' News, 1949, **2**, No 4, 91.

— Thomas Addison: Pioneer of endocrinology. (The Addison Lecture given at Guy's Hospital on 23 June 1949) Brit. med. J. II, 347.

— Medical research as an aim in life. (Address to the Roy. Med. Soc., Edinb.,) Edinb. med. J. **56**, 273.

— Barcroft Memorial Conference. Haemoglobin. (Tribute to Sir Joseph Barcroft), p. 4. London: Butterworths Scientific Publications.

— The atomic problem now. Spectator, 30 September 1949, p. 409.

— Letter of resignation of Honorary Membership of the Soviet Academy of Sciences of the U.S.S.R. Discovery **10**, 32.

1950. — The science of genetics. Caribbean med. J. 1950, **12**, 96.
— Advances in medicinal therapeutics. Brit. Med. J. I, 1.
— Some personal memories of Lord Rutherford of Nelson. Lecture delivered at Cawthron Institute, Nelson, New Zealand. Nelson: Published by R. W. Stiles and Co. Ltd.
— The action and uses of the antihistamine drugs as applied to dermatology. (Introduction to a discussion held at the Annual Meeting of the Brit. Ass. Derm., 1949) Brit. J. Derm. **62**, 151.
— The discovery of insulin, and its introduction into therapeutics. Rapport présenté au II Congrès International de Thérapeutique, June 1949. Editions Arscia, Bruxelles, 1950.
— The pharmacology of histamine: with a brief survey of evidence for its occurrence, liberation, and participation in natural reactions. (Introduction to a Symposium on Antihistamines, October, 1947) Ann. N.Y. Acad. Sci. **50**, 1017.
— Scientific method in medical research. (Opening to a course of lectures: The scientific basis of medicine, arranged by the British Postgraduate Medical Federation) Brit. med. J. II, 1185.
— Science and broadcasting. B.B.C. Quarterly **5**, 136.
— Professor E. D. Adrian, O. M., F. R. S. Electroenceph. clin. Neurophysiol. (Special Number), **2**, 373.
— The chemical transmission of nerve impulses. In: Nervous system. Chambers Encyclopaedia, p. 776. London: George Newnes Ltd. vol. 9,
— Sir Frederick Gowland Hopkins. Dict. of Nat. Biography. Suppl. 1941–1950, p. 406.
— Sir Henry Lyons. Dict. of Nat. Biography. Suppl. 1941–1950, p. 543.
— Rutherford and the Atom. Spectator, 1 December 1950, p. 607.
1951. — Medicine, yesterday and tomorrow. Brit. med. J. II, 1.
— The eye as a physiological reagent. (Bowman Lecture, Ophthal. Soc., 29 March 1951) Trans. ophthal. Soc. U.K. **71**, 117.
— Introduction to Paul Ehrlich, by Marthe Marquardt, 1949, p. 13. London: William Heinemann Medical Books Ltd. New York: Henry Schuman, 1951. (In German). Berlin-Göttingen-Heidelberg: Springer 1951.
— Measurement in medicine. Introduction by Sir Henry Dale. Brit. med. Bull. **7**, 261.
— Harvey and the circulation of the blood. The history of science, p. 55. London: Broadcast Talks, Cohen & West Ltd.
— Transmission of excitation from nerve endings. The evidence for a chemical agency. Times Review of the Progress of Science (Quarterly), No. 2, p. 3.
— The changing face of medicine. Practitioner **167**, 309.
— What Nagasaki means. Spectator, 13 July, 1951, p. 54.
1952. — Mécanisme de l'anaphylaxie. (Fiftieth Anniversary of the Discovery of Anaphylaxis). Presse méd. **60**, 680. Also (in English) Acta allerg. **5**, 191.
— Sir Richard Gregory. (Obituary) Nature (Lond.) **170**, 521.
— Transmission of effects from nerve endings. A survey of the significance of present knowledge. (Lecture at University of St. Andrews Medical School. 13 March 1952). Oxford: Univ. Press.
1953. — The place of research in medical education and practice. (Inaugural Address. Royal Free Hospital School of Medicine. 3 October 1952.) Roy. Free Hosp. Mag. No 16.
— In: What I believe, by Sir James Marchant, p. 67–82. London: Odhams Press Ltd.
— Reizübertragungen durch chemische Mittel im ganzen Nervensystem. Wien. klin. Wschr. **65**, 455.
— Adventures in physiology, with excursions into autopharmacology. London: Pergamon Press.
— A chemical phase in the transmission of nervous effects. Endeavour **12**, No 47, 117.
— Charles Halliley Kellaway, 1889–1952. Obit. Not. Fellows Roy. Soc. **8**, 503.
— The Library of the Royal Society of Medicine. Brit. med. J. II, 1264.
1954. — Paul Ehrlich—Born March 14, 1854. Brit. med. J. I, 659.

1954. — Histamine and its antagonists. J. Lar. Otol. Lond. **68**, 28.
— The Royal Society of Medicine Wellcome Research Library. Proc. roy. Soc. Med. **47**, 1.
— Changes and prospects in Medicinal treatment. The Harben Lectures 1953. J. R. Inst. Pub. Hlth Hyg. **17**, 39, 70, 98.
— The beginnings and the prospects of neurohumoral transmission. Pharmacol. Rev. **6**, 7.
— Dedication of the Charles H. Best Institute. Diabetes **3**, 30.
— PAUL EHRLICH. Clin. nuova (Roma) **18**, 403.
— PAUL EHRLICH and modern therapeutics. Ned. T. Geneesk. **98**, 3028.
— An autumn gleaning. London: Pergamon Press.
— Memorable experiences in research. Diabetes **3**, 20.
— The changing outlook in medicine. Diabetes **3**, 415.
— Hormones of the pituitary posterior lobe. Banting Memorial Lecture. Diabetes **3**, 435.
1955. — Junctional transmission of nervous effects by chemical agents. Mayo Foundation Lecture. Proc. Mayo Clin. **30**, 5.
— EDWARD MELLANBY, 1884–1955. Biog. Mems Fellows Roy. Soc. **1**, 193.
— A historical survey of knowledge of histamine and its functions. Int. Ass. Allergology Congr. Rio de Janeiro, Brazil.
— Humanity's rising debt to medical research. Paget Memorial Lecture. Conquest **44**, 2.
1956. — Medical aims and ideals. The Huxley Lecture. Brit. med. J. **I**, 1125.
— Evidence concerning the endocrine function of the neurohypophysis and its nervous control. Colston Papers **8**, 1.
1957. — Medicinal treatment: its aims and results. Second Gideon de Laune Lecture. Brit. med. J. **II**, 423.
— PERCIVAL HARTLEY, 1881–1957. Biog. Mems Fellows Roy. Soc. **3**, 81.
1958. — Autobiographical sketch. Perspect. Biol. Med. **1**, 125.
— Pharmacy and modern therapeutics. Pharmacol. J. **127**, 292.
— ARTHUR JAMES EWINS, 1882–1958. Biog. Mems Fellows Roy. Soc. **4**, 81.
— Introductory Address to Symposium "Biochemical Aspects of Hypersensitivity". Excerpta Medica Foundation, No 28, C 22.
— Scientific Meeting of the Faculty of Anaesthetists on "Shock and allied phenomena". Opening Speech. Ann. roy. Coll. Surg. **23**, 265.
1959. — A defence of free learning by Lord Beveridge. Review by Sir HENRY DALE in Discovery, Oct. 1959, p. 446.
1960. — The Medical Research Council: A memoir of its beginnings. New Sci. **7**, 998.
— Opening Address at IV European Congr. of Allergy. Acta allerg. (Kbh.) Suppl. VII, p. II.
1961. — THOMAS RENTON ELLIOTT, 1877–1961. Biog. Mems Fellows Roy. Soc. **7**, 53.
— The relation of experimental research to medical knowledge and practice. King's College Hosp. Gaz. **41**, 37.
1961. — Some endocrinological memories. In Perspectives in biology, p. 19. Elsevier: Amsterdam-London-New York.
1962. — OTTO LOEWI, 1873–1961. Biog. Mems Fellows Roy. Soc. **8**, 67.
— Comparative aspects of neurohypophysial morphology and function. Symp. Zool. Soc. Lond. No 9, 179.
1963. — In Memoriam OTTO LOEWI. Ergebn. Physiol. **52**, 1.
— Pharmacology during the past sixty years. Ann. Rev. Pharmacol. **3**, 1.
— Award to Sir HENRY DALE of the Schmiedeberg-Plakette. Naunyn-Schmiedebergs Arch. exp. Path. Pharmak. **246**, 1.
— Medical Research Council Jubilee, 1913–1963: Fifty Years of Medical Research. Brit. med. J. **II**, 1279, 1287.
— Foreword to Supplementary Volume of Heffter's Handbuch der experimentellen Pharmakologie, Bd. 18, S. 26. Berlin-Göttingen-Heidelberg: Springer.
1964. — Sir MICHAEL FOSTER, K.C.B., F.R.S. A Secretary of The Royal Society. In Not. Rec. Roy. Soc. Lond. **19**, 10.

Presynaptic Inhibition in the Vertebrate Central Nervous System

ROBERT F. SCHMIDT* **

With 33 Figures

Table of Contents

* The experiments reported from this laboratory have been supported by the Deutsche Forschungsgemeinschaft.

** II. Physiologisches Institut der Universität Heidelberg, Germany. Present address: Physiologisches Institut der Universität, 2300 Kiel, Germany.

1. Introduction

1.1 Aim and Organization of the Review

A dozen years ago the term ''presynaptic inhibition'' meant little if any-
thing to the majority of neurophysiologists. Meanwhile, intensive investigations
in a considerable number of laboratories have shown that, in the vertebrate
nervous system, presynaptic inhibition is a very powerful and highly organized
type of central inhibition which is particularly potent in controlling the central
excitatory actions of almost all types of primary afferent fibres. Some aspects
of this research have been reviewed previously (ECCLES, 1961a, b, 1963,
1964a, b). This review is an attempt to give a more complete survey of the
tremendous literature which has accumulated during the last 10 years. Special
emphasis has been put on the work which has been published from 1964 up
to and including 1969.

In the introduction the historical development of current concepts on presynaptic inhibition, primary afferent depolarization, and their interrelation
will be described (chapters 1.2 and 1.3). Thereafter, in the first part of the
review (chapters 2.1–2.5), it will be shown that the synaptic mechanism of
presynaptic inhibition is not yet fully understood and that the original hypotheses of ECCLES and his co-workers have not remained unchallenged. But up

to the present time most of the experimental findings support the concept put forward by the Canberra group, and this concept will therefore be used as a framework against which the various experimental results and the conclusions drawn from them will be outlined and evaluated. The central organization of presynaptic inhibition will be dealt with in the second part of the review (chapters 3.1–3.7), and in the last part a preliminary attempt will be made to assess the possible functional significance of this inhibitory mechanism.

Many abbreviations have been used in the text to avoid the frequent repetition of lengthy expressions. A list of the abbreviations will be found on p. 86.

1.2 Current Concepts of Presynaptic Inhibition, Primary Afferent Depolarization, and their Interrelation

The modern history of presynaptic inhibition began in 1957 when Frank and Fuortes described a depression of monosynaptic excitatory potentials (EPSPs) that occurred in the absence of any postsynaptic potential change. There was also no change in motoneuronal excitability, as measured by stimulation of the cell through the intracellular microelectrode, and the antidromic invasion was neither helped nor hindered, as was shown when the safety factor for invasion was reduced by hyperpolarizing currents. Two alternative explanations were offered by Frank (1959) for this finding: either the postsynaptic inhibitory changes took place far out on the dendrites of the motoneurone and could not be "seen" (quotation marks used by Frank, 1959) by the intracellular microelectrode, or the inhibitory nerve impulses interacted with the excitatory volley before the latter had arrived at the surface of the motoneurone, and blocked transmission in the terminal fibres. Since both of these explanations would locate inhibitory action at a site remote from the motoneuronal soma, Frank (1959) designated the phenomenon "remote inhibition".

It is most unfortunate that the pioneering results of Frank and Fuortes (1957) have never been published in detail. Eccles, Eccles and Magni (1961) added the observation that during "remote" inhibition there was no detectable change in the time course of the depressed EPSPs (Fig. 1). They concluded that this finding and those of Frank and Fuortes (1957) were indicative of a presynaptic rather than a dendritic site of action. At the same time it was shown that the inhibition was associated with a depolarization of the afferent fibres (see below), and, therefore, Eccles, Eccles and Magni (1960, 1961) designated this inhibitory phenomenon as "presynaptic inhibition", which was the term originally used by Frank and Fuortes (1957).

The postsynaptic membrane properties during presynaptic inhibition were re-investigated by Eide, Jurna and Lundberg (1968). These authors presented additional evidence that during presynaptic inhibition of EPSPs induced by Ia afferents no conductance change of the motoneuronal membrane could be detected. Furthermore, they confirmed that the time course of these EPSPs did not change during the depression. They concluded that the unchanged rate of decay, which should be particularly sensitive to dendritic conductance changes (RALL, 1962, 1964; BURKE, 1967), excludes the possibility that the EPSP depression is caused by remote postsynaptic inhibition on dendrites. Finally, they showed that during presynaptic inhibition of the Ia

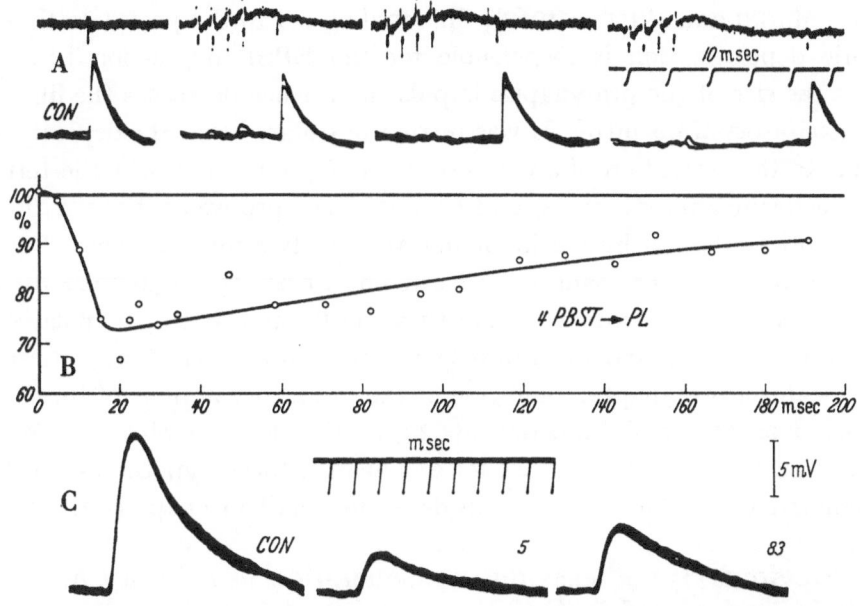

Fig. 1 A–C. Depression of monosynaptic EPSP by presynaptic inhibition. In A the EPSP (CON) in a plantaris motoneurone is seen to be depressed by four Group I conditioning volleys in the nerve to the knee flexors, posterior biceps plus semitendinosus (PBST). The timing of the conditioning and testing afferent volleys is shown in the upper traces (positivity upwards in both traces). In B the time course of the EPSP depression (expressed as percentage of control) is shown for the series illustrated in A. In C the control EPSP (CON) of another experiment is seen to be greatly depressed both at 5 and 83 ms after a conditioning tetanus of 22 Group I volleys (ECCLES, ECCLES and MAGNI, 1961)

induced EPSPs the size and the time course of the EPSPs generated monosynaptically by volleys in descending tracks of the spinal cord were unaltered. Thus it appears established beyond reasonable doubt that monosynaptic Ia induced EPSPs can be depressed by a presynaptic inhibitory mechanism.

It should be added that in a number of papers GRANIT and co-workers have shown that muscle stretch often produced EPSP depressions without membrane hyperpolarizations but at the same time there was a reduction of the motoneurone discharge set up by constant transmembrane currents (GRANIT et al., 1964; GREEN and KELLERTH, 1966; GRANIT, 1968; KELLERTH, 1968). They concluded that under these experimental conditions presynaptic inhibition was either weak or absent, and that remote dendritic inhibition was possibly of greater physiological significance than presynaptic inhibition. Since muscle stretch produces activity in a great variety of afferent fibres, it appears doubtful whether the results obtained in these complex experimental situations permit estimations of the relative importance and effectiveness of either remote dendritic or presynaptic inhibition.

Eccles and his co-workers observed that the presynaptic inhibition of Ia induced EPSPs precisely correlated with a depolarization of presynaptic fibres (primary afferent depolarization, PAD: Eccles, Eccles and Magni, 1961; Eccles, Magni and Willis, 1962). This finding led to the postulate that presynaptic depolarization is responsible for the EPSP depression because it reduces the size of the presynaptic impulse and hence decreases the liberation of the excitatory transmitter. It was further postulated that the depolarization was due to the activation of an axo-axonic synapse located near the terminal of the recipient afferent fibres, and that this synapse was activated through polysynaptic reflex pathways involving at least two interneurones. Furthermore, the observations made on Ia afferent fibres were extended to other types of spinal primary afferents and it was postulated that the depolarization of Ib afferent fibres and of Group II and III cutaneous afferent terminals also indicates a presynaptic inhibition of the excitatory actions of these fibres (Eccles, Kostyuk and Schmidt, 1962a, b; Eccles, Schmidt and Willis, 1963a). The considerable evidence for and against these hypotheses which has accumulated meanwhile will be considered in detail in chapters 2.1–2.5.

1.3 Previous Hypotheses on the Generation of Primary Afferent Depolarization and its Relation to Central Inhibitory Processes

It should first be recognized that during the last 35 years a variety of hypotheses has been put forward to explain the mechanism of presynaptic depolarization and its possible correlation with inhibitory processes. Initially it was postulated by Barron and Matthews (1935, 1938) that the prolonged (200 ms) positive potential (P-wave) which could be recorded from the surface of the cord dorsum following an afferent volley (Gasser and Graham, 1933) was produced by the same potential generator as the negative dorsal root potential (DRP) led off from a dorsal rootlet. It was concluded that these potentials reflected a prolonged depolarization of the dorsal root fibres. This identification has been accepted by all subsequent investigators. However, there was much diversity in the attempts to explain the manner in which DRPs were produced. Barron and Matthews (1938), Dun (1941) and Brooks and Fuortes (1952) suggested that DRPs arise primarily in dorsal root fibres by ionic changes external to their terminals (for instance by an accumulation of potassium in the extracellular space during activation of afferent fibres, cf. Trachtenberg and Pollen, 1970) and "by a mechanism analogous to that responsible to the afterpotentials in nerve". According to Bonnet and

BREMER (1938, 1952), ECCLES (1939), BREMER and BONNET (1942, 1949), ECCLES and MALCOLM (1946), LLOYD and McINTYRE (1949), GRUNDFEST and MAGNES (1951) and KOSTYUK (1956) the primary afferent fibres do not act directly on each other, but interneurones are the primary generators of the DRP, there being some unspecified mechanism of electrical transmission from them to the terminals of the primary afferent fibres that are in close apposition. It seemed impossible otherwise to explain the very wide segmental distribution of the depolarization, and particularly the potentials in the contralateral dorsal roots.

A further argument against the direct interaction of primary afferent terminals was based on the observation that impulses in a particular afferent fibre had no specific action in depolarizing that fibre (KOKETSU, 1956 a, b; ECCLES and KRNJEVIĆ, 1959a), and that volleys in one dorsal root produced in that root a DRP that was about the same size as the DRP produced by an adjacent root (BARRON and MATTHEWS, 1938; ECCLES and MALCOLM, 1946; LLOYD and McINTYRE, 1949).

A possible relation between primary afferent depolarization and central inhibitory processes was first suspected by GASSER and GRAHAM (1933), when comparing the slow positive potential waves recorded from the cord dorsum (P-wave), and the inhibition of flexor reflexes after single ipsilateral conditioning volleys in peripheral nerves (FORBES et al., 1928; GERARD and FORBES, 1928; ECCLES and SHERRINGTON, 1931). The inhibition as well as the P-wave reached a maximum after about 20 ms and persisted for as long as 400 ms (compare also Fig. 1 B). In view of the similarity of time courses, GASSER and GRAHAM (1933) asked whether the positive potential may not be connected with the process responsible for inhibition. In a further investigation, HUGHES and GASSER (1934) provided additional evidence that the size and time course of the P-wave produced by a conditioning volley could be correlated with the size and time course of the depression both of the N-waves on the cord dorsum, and of the reflexes that are produced by a subsequent testing volley. BARRON and MATTHEWS (1938), in agreement with GASSER and collaborators, also postulated that the potential generator producing DRP and P-wave was responsible for inhibition. They attributed the inhibition to electrical currents which caused blockage of conduction in collateral branches of interneurones, whereas GASSER (1937) proposed that depression of interneurones was due to positive after-potentials following their spike discharge. It was not shown how this positive after-potential could give rise to the P-wave and the DRP.

Later on, the evidence relating both the DRP and the P-wave to a central inhibitory action has been lost sight of, because the interneuronal theory of inhibition (GASSER, 1937; BONNET and BREMER, 1939; BREMER and BONNET, 1942) could not explain inhibition of monosynaptic reflexes (LLOYD, 1941, 1946; RENSHAW, 1941, 1942), and also because central inhibition was attributed to the postsynaptic action of special inhibitory synapses (see Eccles, 1953, 1957, 1964 for reviews). It was generally assumed that there was merely an incidental relationship between DRP and P-wave on the one hand, and the postsynaptic inhibitory action of interneurones on the other. As a consequence, it could be stated that the DRP was of "no functional significance" (Fatt, 1954). The work of Eccles and his colleagues has re-established the association of primary afferent depolarization with reflex depression. Their detailed postulates, both on the mode of generation of the PAD and on the mechanism of the resulting inhibition, provoked a new interest in this important field of central sensory physiology.

2. Mechanism of Presynaptic Inhibition in the Vertebrate Central Nervous System

2.1 Electrophysiological Manifestations of Primary Afferent Depolarization and the Methods for their Detection

Fig. 2 shows schematically the principal methods used in the investigation of primary afferent depolarization. Not included in the figure is the recording of slow potentials inside the spinal cord with an extracellular microelectrode. Each of the methods shown has its particular potentialities and limitations and these will be outlined briefly. All of the techniques are able to detect the occurrence of PAD, but they differ greatly in their capability to signal what types of fibres are depolarized. Two of them, the recording of dorsal root

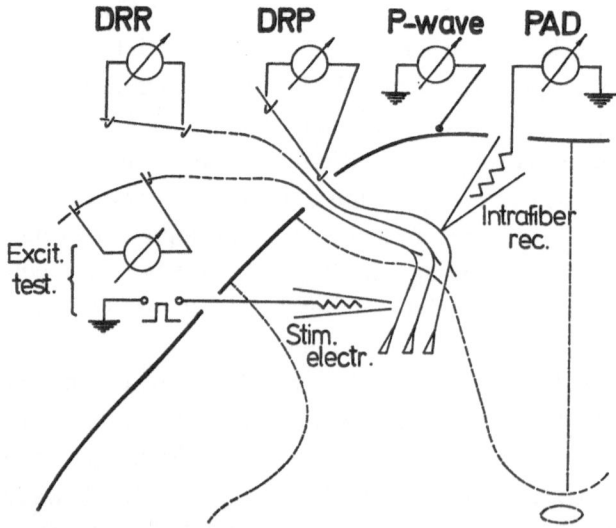

Fig. 2. Methods for detecting primary afferent depolarization, PAD. From right to left these methods are: The intrafibre recording of the PAD from primary afferents with glass capillary microelectrodes; the recording of the positive potential wave, P-wave, from the dorsum of the spinal cord near the dorsal root entry zone; the recording of dorsal root potentials, DRP, with bipolar Pt or Ag/AgCl wire electrodes from dorsal rootlets cut off a few mm distal to their dorsal root entry; the recording of dorsal root reflexes, DRR, either in dorsal root filaments, or preferentially in peripheral nerves; and the testing of the excitability of the terminal sections of primary afferent fibres

potentials (DRP) and the recording of slow positive potentials from the cord dorsum (P-waves) give no hint at all as to the fibre types receiving the depolarization, while two others, the recording of dorsal root reflexes (DRR) and the testing of intraspinal excitability changes of fibre populations, allow the classification of the depolarized fibres in regard to their conduction velocity and their origin from muscle or cutaneous nerves. But only the most sophisticated ones, intracellular recording and testing of the excitability of single afferent fibres, can be used for the exact determination of the modality to which the fibre under observation belongs.

2.1.1 Intrafibre Recording

Intracellular recording from primary afferent fibres in the dorsal horn of the spinal cord is the only direct method for measuring potential changes which occur across the membrane of these fibres either during their own activation or during the activation of other primary afferent fibres (Fig. 3;

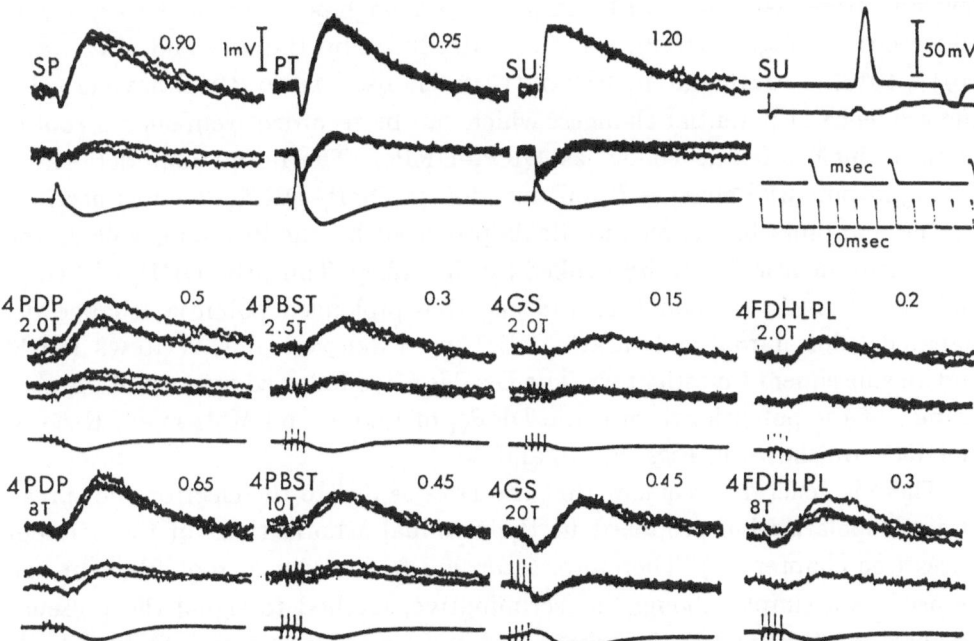

Fig. 3. Intracellular recording of the primary afferent depolarization (PAD) of a fiber of the SU nerve at 0.7 mm depth. The SU record in the upper right-hand corner shows in the upper trace the intracellular spike and in the lower the cord dorsum potential. All other records were at a higher amplification and at a much slower sweep speed in order to display the depolarizations produced by afferent volleys in various cutaneous and muscle nerves of the hind limb, as indicated by the symbols. Upper traces are the intracellular records, depolarization being upward. Middle traces show the field potentials similarly recorded, but with the microelectrode just outside the fiber; the lower traces show the cord dorsum potentials, but with upward deflexion negative. All records are formed by the superposition of several traces, usually three. Subtraction of the extracellular fields from the intracellular potentials gives the PAD, which are shown in mV for each record. The cutaneous nerves in the first row have been stimulated with single shocks of four times threshold strength. All other nerves were stimulated with four shocks at 300/s and the stimulus strength is indicated on each record relative to the threshold (T). The 1 mV calibration is for the intracellular and extracellular records. The 10 ms time marker is for all except the spike record, which was recorded at the faster sweep speed (ECCLES, SCHMIDT and WILLIS, 1963e)

see also Figs. 4, 23 and 25). The potentials recorded just outside the primary afferent fibres (middle traces in Fig. 3) have to be subtracted from the intracellular recordings (upper traces in Fig. 3) to obtain the actual membrane potential changes. At the segmental level, stable intrafibre recordings with normal resting membrane potentials can only be expected in the dorsal part of the spinal cord, where the afferent fibres have their largest diameter. Thus, it is usually impossible to record intracellularly from preterminal sections of the afferent fibres, except perhaps in the dorsal funiculus nuclei, where the afferent fibres seem to expand at their ultimate terminals (Walberg, 1965), and where physiological evidence indicates that intrafibre records can be obtained within less than 300 μ from the synaptic knobs (Andersen, Eccles, Schmidt and Yokota, 1964b).

2.1.2 Dorsal Root Potentials

The recording of potential changes from a dorsal root or rootlet, cut a few mm from its spinal cord entry and put on bipolar Pt or Ag/AgCl wire electrodes (cf. Fig. 2, DRP), was first introduced by Barron and Matthews (1935, 1938). Lloyd and McIntyre (1949, cf. also Lloyd, 1952) have labelled the sequence of potential changes, which can be recorded from such a rootlet upon a single afferent volley, as DRP_I–DRP_{VI}. The initial four deflections are rapid potential changes (total duration of DRP_I–DRP_{IV} 2–4 ms) ascribed by these authors to the electric fields produced by the incoming volley, and by the interneuronal activity evoked by the volley. Thus, the DRP_V of Lloyd and McIntyre corresponds to the negative prolonged potential change designated "dorsal root potential", DRP, by Barron and Matthews (1938) and all subsequent investigators. The DRP is often followed by a small "underswing" of the potential record, the DRP_{VI} of Lloyd and McIntyre. Records of DRPs are shown in Figs. 15, 19 and 20.

There is general agreement that the DRP is due to the electrotonic spread of the depolarization produced in the terminal arborizations of the afferent fibres (see chapter 1.3). Therefore, recording from a dorsal root filament can be used as a simple, though indiscriminative, method to signal the presence or absence of PAD (see for instance Andersen, Eccles and Sears, 1962, 1964a; Eccles, Schmidt and Willis, 1963a, b, c, e, f; Cook et al., 1969a; Eisenmann and Rudin, 1954; Fetz, 1968; Franz and Iggo, 1968; Jänig et al., 1967; Mallart, 1965; Mendell and Wall, 1964; Schmidt, 1966; Schmidt and Willis, 1963b; Tang, 1969; Zimmermann, 1968; Chu, 1970). In the amphibian spinal cord, especially large DRPs can be recorded both *in vivo* (Barron and Matthews, 1938; Bonnett and Bremer, 1948, 1952; Fuortes, 1951) and *in vitro* (Eccles and Malcolm, 1946; Malcolm, 1953; Takagi, 1951; Göpfert, 1956). The latter preparation has proved to be particularly suitable for the investigation of the influence of ions and drugs on PAD (Eccles and

MALCOLM, 1946; ECCLES, 1947; TAKAGI, 1954; KIRALY and PHILLIS, 1961; SCHMIDT, 1963; RICHENS, 1969; TEBĒCIS and PHILLIS, 1969a, b; TEBĒCIS, 1970).

2.1.3 Dorsal Root Reflexes

Highly synchronized afferent volleys, such as those evoked by electrical stimulation of peripheral nerves, or by brief and intense mechanical stimuli (e.g. tapping the foot pads or the tibia), often produce PAD with a rising phase steep enough to elicit antidromic action potentials (Fig. 4E–G). These

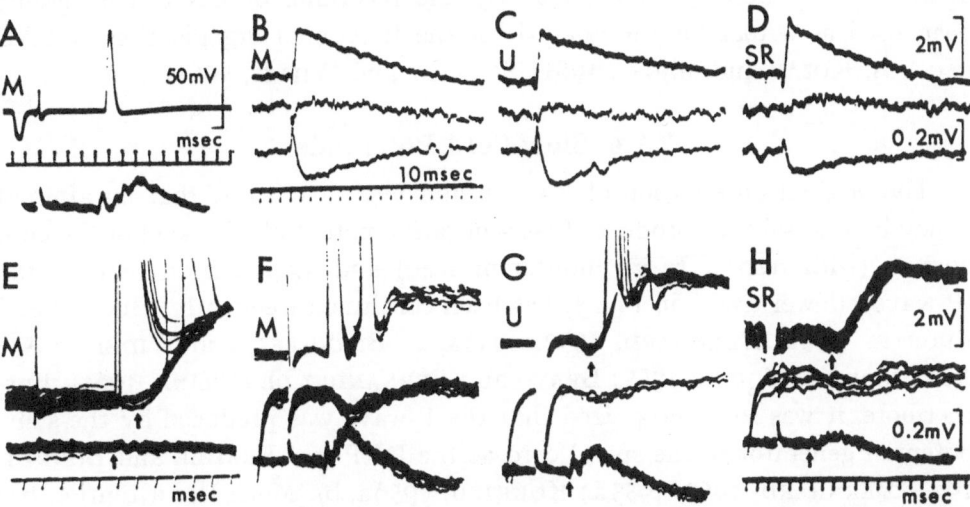

Fig. 4A–H. Intracellular recording of dorsal root reflexes. A Identification of the fibre of the median nerve (*M*) in the cuneate nucleus. B–D Primary afferent depolarizations (*PAD*) in response to single volleys in the median (*M*), ulnar (*U*), and superficial radial nerve (*SR*) respectively. E Spike and onset of *PAD* with dorsal root reflexes (*DRR*) taken at fast sweep speed. Since these *DRR* were generated in the cuneate nucleus they were termed dorsal column reflexes (*DCR*). F–H Spikes, *PAD* and *DCR* to single *M*, *U* and *SR* volleys. Latency of *PAD* onset indicated as the time between the first spike in surface record and the second arrow (ANDERSEN, ECCLES, SCHMIDT and YOKOTA, 1964b)

antidromic discharges can be recorded in peripheral cutaneous and muscle nerves as "dorsal root reflexes", DRR. Thus the occurrence of DRRs signals the presence of a large and steeply rising PAD in afferent fibres of the nerve under observation. If the PAD falls below a critical threshold the DRR disappears. It is therefore not surprising that inhibition of PAD is always accompanied by a depression of the DRRs which is more spectacular than that of the DRP (cf. ECCLES, KOZAK and MAGNI, 1961; SCHMIDT and WILLIS, 1963 b).

DRRs were first seen in dorsal roots by MATTHEWS (1934) and in peripheral cutaneous nerves by TOENNIES (1938, 1939). Later on BROOKS et al. (1955) and BROOKS and KOIZUMI (1956) demonstrated that DRR can also be recorded in muscle nerves, particularly in the cooled animal, where in a manner not yet understood lower thresholds are found for reflex discharges along both ventral and dorsal roots. The properties of DRR both in the cat and the amphibian spinal cord have been investigated in some detail. It was recognized that the DRP and DRR had a common origin and several explanations of their interconnection have been suggested (Matthews, 1934; Barron and Matthews, 1935, 1938; Toennies, 1938, 1939; Dun, 1939; Barron, 1940; Skoglund and Uvnäs, 1943; Eccles and Malcolm, 1946; Bonnet and Bremer, 1952; Habgood, 1953; Brooks and Koizumi, 1956; Göpfert, 1956; McCouch and Austin, 1958; Tregear, 1958; Calma and Quayle, 1968; Tebēcis and Phillis, 1968). Recently the recording of DRRs has mainly been used as a tool for the analysis of the functional organization of PAD (Eccles, Kozak and Magni, 1961; Schmidt and Willis, 1963b).

2.1.4 Slow Cord Potentials

The original observation of Gasser and Graham (1933) that an afferent volley in a dorsal root produces first a negative potential (N-wave) of the cord dorsum (with respect to an indifferent lead) and later a positive potential (P-wave), (lower traces in Fig. 3) has been confirmed frequently (Brooks and Fuortes, 1952; Bernhard, 1952, 1953a, b; Bernhard and Widén, 1953; Austin and McCouch, 1955; Bravo and Fernandez de Molina, 1961). Furthermore, it was soon recognized that the P-wave was produced by the same potential generator in the spinal cord as the DRP (see Bremer and Bonnet, 1942; Bernhard, 1952, 1953a; Koketsu, 1956a, b). Much less attention has been paid to the potential fields within the spinal cord which correspond to the P-waves (Koketsu, 1956b; Coombs et al., 1956), and which have to be measured after the end of the N-waves, i.e. approximately 40 ms after the afferent volley entered the cord, so as to avoid interference by the initial waves. The distribution of the sinks and sources at this time gave valuable clues as to the location of the afferent fibre sites acting as potential generators, provided that the afferent volley was produced by the stimulation of specific fibre types (cf. Figs. 10 and 12; see also Eccles, Magni and Willis, 1962; Eccles, Kostyuk and Schmidt, 1962a; Eccles, Schmidt and Willis, 1963a). The interpretation of intraspinal potential fields, measured after stimulation of dorsal roots (cf. Howland et al., 1955; Wall, 1962), is more difficult since dorsal root stimulation results in activity in all types of myelinated afferent fibres.

2.1.5 Excitability Testing

Wall (1958) has presented convincing evidence that under favourable conditions of population distribution of primary afferent fibres in proximity to an intraspinal stimulating electrode, the number of primary afferent fibres excited antidromically by a test stimulus was approximately proportional to the strength of the stimulus; and hence that the size of the antidromic mass discharge recorded in the peripheral nerves was a measure of the excitability of the central fibre segments. Furthermore, he concluded that under such conditions changes in excitability indicated a change of the resting membrane potential of the population of fibres under observation, increased excitability signalling depolarization, and vice versa.

Meanwhile the method has proved to be a most valuable tool for the investigation both of the synaptic mechanism and the organization of presynaptic inhibition in the spinal cord, the dorsal column nuclei, and in various

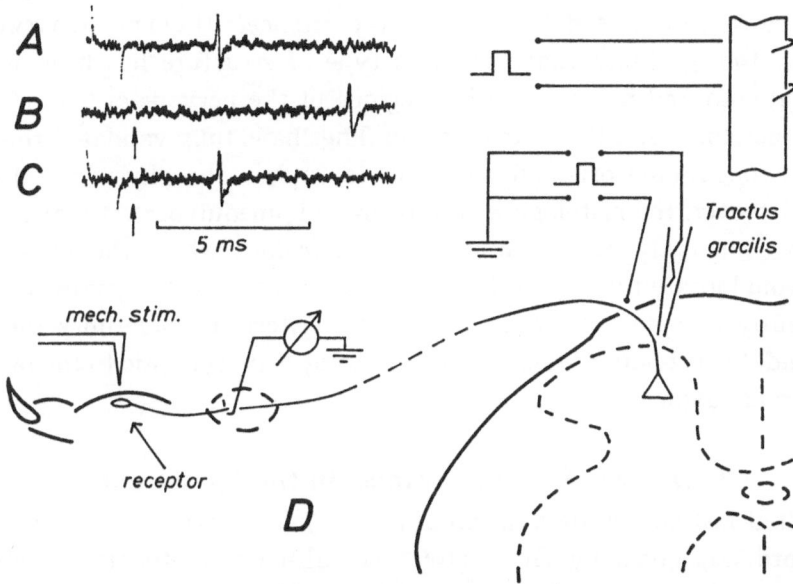

Fig. 5 A–D. Excitability testing of the terminal sections of single primary afferents coming from identified cutaneous mechanoreceptors. D Schematic drawing of the arrangement of the stimulating and recording electrodes. A–C Specimen records of action potentials recorded from a fine filament of the N. plantaris. A After peripheral activation of a mechanoreceptor. B Following stimulation of the same afferent fibre in the spinal cord through a platinum wire microelectrode. C following simultaneous peripheral and central stimulation. Collision prevents the appearance of the centrally evoked action potential. The central stimulus is indicated by arrows (SCHMIDT, SENGES and ZIMMERMANN, 1967a)

nuclei of cranial nerves. Since the testing and recording conditions remain virtually constant for very long periods of time (up to many hours), a great variety of experimental situations can be tested with respect to changes in the excitability of well defined populations of afferent fibres (see for instance ECCLES, MAGNI and WILLIS, 1962; ECCLES, SCHMIDT and WILLIS, 1963 a, b, e; ANDERSEN, ECCLES, SCHMIDT and YOKOTA, 1964b; DARIAN-SMITH, 1965). Furthermore, a modification of this technique enables the investigator to determine the excitability changes of single afferent fibres in the spinal cord, and to identify simultaneously the receptor properties of the unit under observation (Fig. 5). This method is particularly helpful when probing into the functional organization of the presynaptic inhibitory pathways, by using adequate peripheral stimulation (Schmidt, Senges and Zimmermann, 1967a, b; Jänig, Schmidt and Zimmermann, 1968b; Schmidt, 1968, 1969; Vycklický et al., 1969).

2.2 The Histology of Axo-Axonic Synapses

When it was first proposed that primary afferent depolarization was due to the action of a chemically operated synapse located near the presynaptic terminals of primary afferent fibres (Eccles, 1961b), it was pointed out that "no histological evidence can be adduced in respect either of the location of these synapses, or indeed for their very existence" (Eccles, Kostyuk and Schmidt, 1962a). Since that time this type of structure has been found in the spinal cord and in several other regions of the nervous system. It might appear, therefore, that the histological findings have fully validated this model that was constructed originally in order to explain the physiological experiments. However, the histological results are not unequivocal: the presynaptic synapses cannot always be found in those regions where the physiological results would locate them, and the functional polarity of the synapse as judged by presently accepted electronmicroscopical criteria is sometimes the wrong way round, the primary afferent fibre appearing as presynaptic to the presumed interneuronal axon.

2.2.1 Axo-Axonic Contacts in the Spinal Cord

The first histological description of axo-axonic contacts in the mammalian spinal cord was given by Gray (1962, see also Gray, 1963) who observed that boutons in the spinal cord showed membrane thickenings at opposed regions. The specialized region occurred where the small bouton was in contact with a large one. Gray presumed that these contacts form the morphological basis for presynaptic inhibition, although with his preparation it was not possible to determine whether the "postsynaptic" site of the axo-axonic contacts was an afferent terminal.

Meanwhile axo-axonic contacts have been seen in the dorsal regions of the spinal gray matter (Ralston, 1965; Kerr, 1966), and the investigations of Scheibel and Scheibel (1968) and Réthelyi and Szentágothai (1969) have made it likely that such contacts exist between primary afferent collaterals and axons originating from interneurones of the spinal cord. As diagrammatically shown in Fig. 6, Scheibel and Scheibel (1968) have concluded from their Golgi material that the terminal field (n) of the large myelinated afferent fibres (t_1) in laminae II and III (lamina II of Rexed 1952, 1954, corresponds to the Substantia gelatinosa Rolandi) is interpenetrated at the base and apex by secondary axonal systems, which effect complex axo-axonic synapses at each site. Each apex appears to be sheathed in a capping plexus derived largely from gelatinosal cell axons (t_3 in Fig. 6) projecting short distances, rostral or caudal, via Lissauer's tract. Infiltration at the base of each field seems to be effected by terminals (t_2) emerging from the ventro-medial part of the posterior funiculus. The authors assume that these latter fibres represent a combination of propriospinal axons from laminae IV and V neurons, and possibly from primary afferent collaterals from dorsal roots. More recently SCHEIBEL and SCHEIBEL (1969) have postulated that PAD may at least partly be due to direct interfibre interactions in the dense aggregates (microbundles) of dorsal root fibres.

In a combined Golgi and electronmicroscopic study, RÉTHELYI and SZENTÁ-GOTHAI (1969) have tried to elucidate in more detail the structural arrangement and synaptic connections of the various components present in lamina II, and

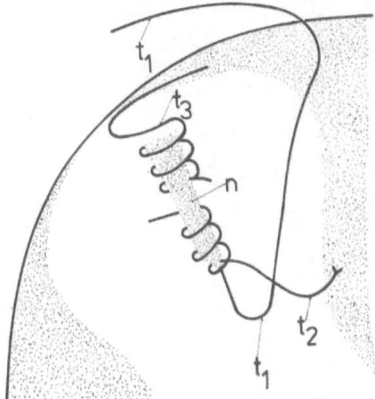

Fig. 6. Schematic drawing of axo-axonal arrangements in the dorsal horn as seen in Golgi-preparations. The primary afferent t_1 generates a terminal neuropil plexus n. Terminal t_2 develops axo-axonic relations with the base of the plexus while terminal t_3, emerging from the area of Lissauer's tract develops axo-axonic relations with the apex of the plexus (SCHEIBEL and SCHEIBEL, 1968)

Fig. 7 gives the schematic diagram of their principal findings. The most conspicuous structures in the Substantia gelatinosa are the large sinusoid axon terminals, which form the centers of large glomerulus-like synaptic complexes (RALSTON, 1965; RÉTHELYI and SZENTÁGOTHAI, 1969; KERR, 1966). As indicated in Fig. 7, the large terminals (labelled DSA in the inset) were shown to be of intraspinal origin, and are thought to arise from pyramid-shaped cells situated at the border between laminae III and IV. [Initially it had been suspected that the large terminals (DSA) arose from primary afferents, SZENTÁGOTHAI, 1968]. The sinusoid axon terminals (DSA) are shown to establish axo-dendritic synapses with lamina II neurones (cells *1* and *2* in Fig. 7), and numerous axo-axonic synapses with smaller terminals (labelled *ST* in the inset). In degeneration experiments, these latter terminals were identified as coming mainly from collaterals of myelinated afferents (fibre *A*) that penetrated into the ventral parts of the dorsal horn and curved back in dorsal

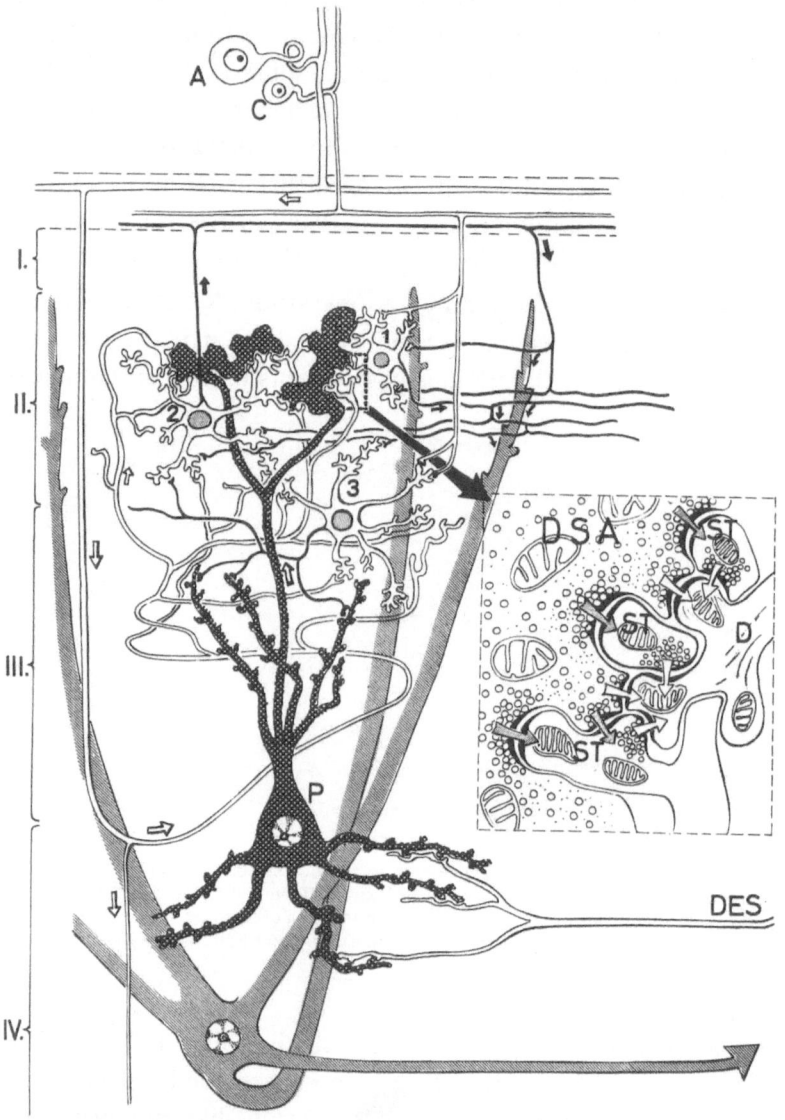

Fig. 7. Diagram illustrating neuronal arrangements in the Substantia gelatinosa and the various elements connected in the synaptic complexes (in dashed square and EM diagram at right). Lamination of Rexed (1952, 1954) is indicated at left margin. The diagram represents a longitudinal section of the dorsal horn. Two primary sensory afferents are indicated: a large calibre (probably cutaneous) afferent, *A*, and an unmyelinated small calibered afferent, *C*. They establish contacts with three SG neurons (*1–3*) two of which belong to lamina II, whereas the third (*3*) is somewhat larger and is situated (partly) in lamina III. Pyramidal cell (*P*, cross hatched) situated at the border of lamina III and IV receives contacts from interneuron (*3*) and possibly also directly from large primary afferent (*A*). The dorsally directed short axon of the pyramidal neuron participates in the SG synaptic complexes. It is assumed that only *A* fibres participate in the complexes but this is not certain. Forward conduction from the SG is secured by large lamina IV neuron (hatched), the dendrites of which are embedded into the longitudinal axonal plexus of the SG (in right upper part of diagram). EM diagram at right shows ultrastructural arrangement in part of a synaptic complex. Central dense sinuous axon terminal (*DSA*) is in contact with dendritic ends (*D*) of substantia gelatinosa neurons and presynaptic to smaller axon terminals (*ST*) that are mainly primary afferent endings (although some might be also axon terminals of SG neurons). White arrows indicate in this part of the figure ordinary synapses and hatched arrows axo-axonic synapses. Direction of the arrows shows the polarity of the synapse as it would appear from structural criteria (Réthelyi and Szentágothai, 1969)

direction. Thus the findings of RÉTHELYI and SZENTÁGOTHAI largely support the conclusions made in the physiological study on the pathway of PAD of large myelinated afferents (ECCLES, KOSTYUK and SCHMIDT, 1962a, see also chapter 2.4 of this review).

Until recently there was much less convincing histological support for the presence of specific axo-axonic contacts on the presynaptic contacts of I a afferent fibres impinging upon motoneurones, although the physiological experiments indicated that the I a fibres are depolarized at these sites during presynaptic inhibition. Nevertheless, axo-axonic contacts were seen, with small axon terminals being impressed into larger synaptic knobs (Fig. 4 in SZENTÁGOTHAI, 1968; KHATTAB, 1968). It had been proposed (PORITSKY, 1969) that the paucity of such findings is due to the fact that the majority of synaptic boutons on motoneurones are not I a terminals, and it had also been suggested (SZENTÁGOTHAI, 1968) that the morphological substrates responsible for the depolarization of I a fibres are situated more remotely from the synapse, for instance in laminae V and VI or at the border between lamina VII and the motor nucleus, where the I a fibres show numerous side branches which would be favourable sites for such contacts. However, a recent study of the ultrastructure of dorsal root boutons on motoneurones (CONRADI, 1969) revealed that on the convex side of these boutons there were regularly apposed small boutons containing synaptic vesicles of irregular shape. These latter boutons established a synaptic complex with the big boutons of dorsal root (Group I a) fibres. A similar arrangement can be found on the I a fibres terminating at Clarke's column neurones (RÉTHELYI, 1970).

2.2.2 Axo-Axonic Contacts in Dorsal Column Nuclei

In the cuneate nucleus, WALBERG (1965) found axo-axonic synapses in configurations corresponding exactly to those predicted on the basis of physiological observations (ANDERSEN, ECCLES, SCHMIDT and YOKOTA, 1964a, b, c). A specimen electron micrograph is shown in Fig. 8, and a diagram summarizing the findings is given in Fig. 9. The presynaptic bouton participating in the axo-axonic contact was usually small, the other one large. Degeneration experiments established that the large boutons were axon terminals of cuneate tract fibres. Furthermore evidence was presented that the small presynaptic boutons did not originate from neurones located outside the cuneate nucleus. This is again in agreement with the physiological findings, as can be seen by comparing WALBERG's diagram (Fig. 9) with those drawn according to the physiological results (Fig. 31).

It should be pointed out that WALBERG (1965), as well as other histologists, observed that the presynaptic axon of the axo-axonic contacts often also formed axo-dendritic contacts on the same dendrites as the postsynaptic axon (axon a_1 in Fig. 9, see also inset in Fig. 7). These arrangements suggest that a

3*

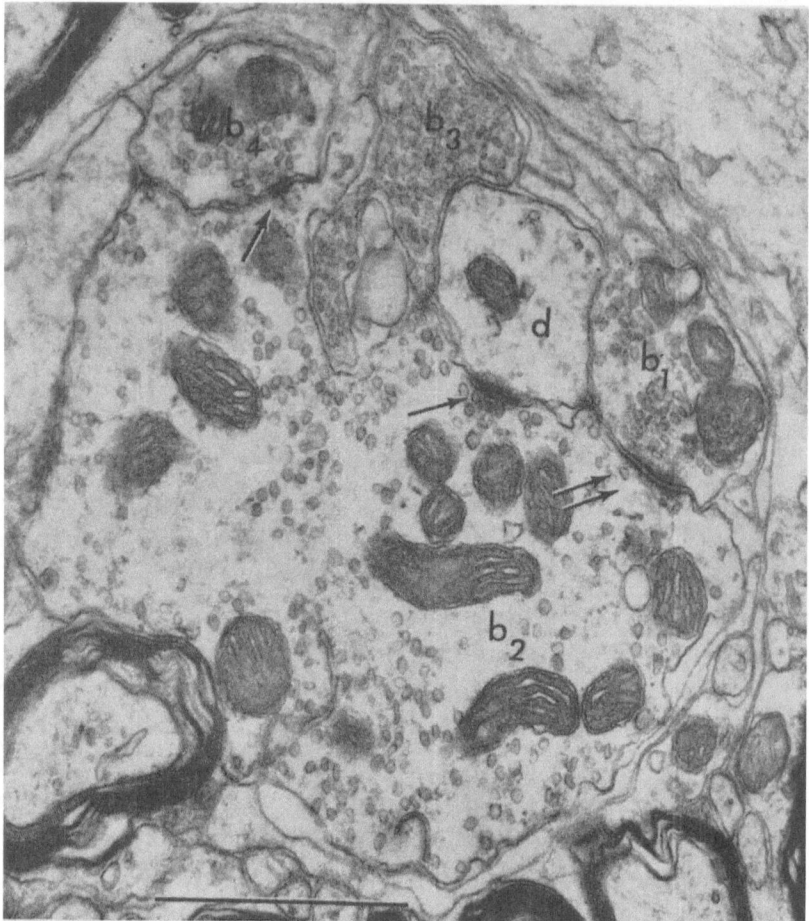

Fig. 8. Axo-axonic contacts in the cuneate nucleus. Electron micrograph from a region of the right cuneate nucleus. The small dendritic branch d is surrounded by several profiles, of which three are boutons (b_1, b_2, b_3). The large bouton b_2 is at the arrow in contact with dendrite d; this is an axo-dendritic synapse. The small bouton b_1 is at the two arrows in contact with the large bouton b_2, and there is here a membrane thickening of both boutons. Synaptic vesicles are accumulated towards the surface membrane of the small bouton. This contact is interpreted as an axo-axonic synapse. Another not conspicuous membrane thickening occurs at the arrow in the small bouton b_4 and the large bouton b_2 where these are in contact (Walberg, 1965)

combined presynaptic and postsynaptic inhibition can be exerted by these axons. Up to date physiological evidence for such a combined inhibition is lacking. (At the crayfish neuromuscular junction, activation of the inhibitory nerve fibres always produces pre- and postsynaptic inhibition, Dudel and Kuffler, 1961).

2.2.3 Axo-Axonic Contacts in Nuclei of the Cranial Nerves and in the Thalamus

Shortly after the first observation of PAD in the trigeminal brain stem nuclei (Darian-Smith, 1965), the presence of axo-axonic contacts in this region has been described by Kerr (1966). These axo-axonic contacts were seen in all layers of the nucleus, and it was clearly established (KERR, 1970) that in the majority of cases the primary afferent fibres were postsynaptic to presynaptic knobs of unknown origin containing flattened vesicles. The axo-axonic contacts of trigeminal afferents in the Substantia gelatinosa associated with the spinal tract of the trigeminal nerve probably correspond to those described by

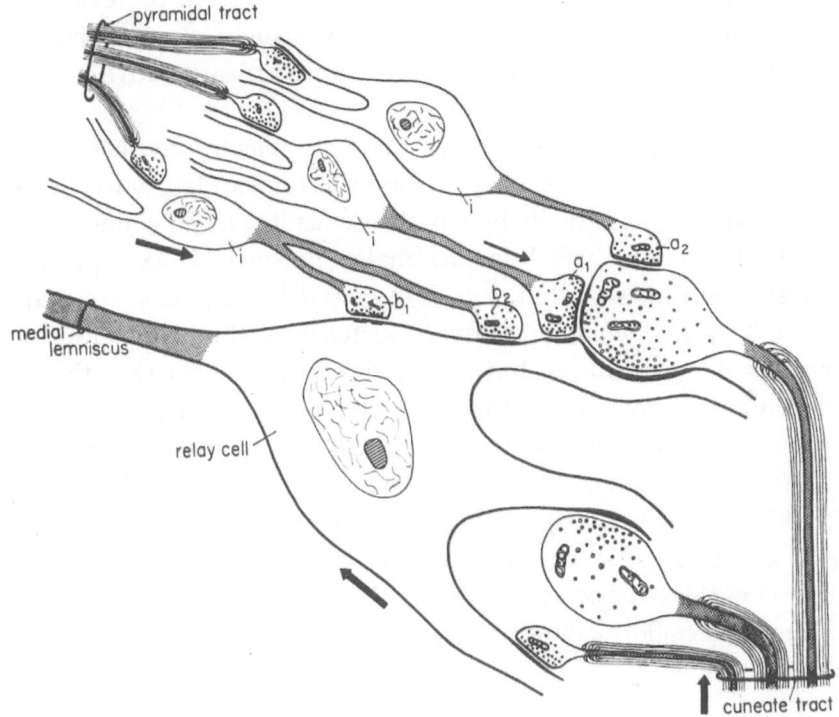

Fig. 9. Diagram showing the pattern of synaptic contacts in the cuneate nucleus. The large and probably also some small axon terminals of the fibres of the cuneate fascicle make contact with dendrites of relay cells. The small terminals of the pyramidal tract fibres have synapses with dendrites of interneurons (i). The axon terminals of the interneurons synapse with soma and dendrites of relay cells (b_1 and b_2). Some of these terminals may also be derived from cells of the adjacent reticular formation. Some of the axon terminals of interneurons which make axo-dendritic synapses are in addition in contact with large axon terminals of fibres in the cuneate fascicle. This is an axo-axonic synapse (a_2). Other terminals of the interneurons are in contact only with terminals of the fibres of the cuneate fascicle: this axo-axonic synapse is shown at a_1 (WALBERG, 1965)

RÉTHELYI and SZENTÁGOTHAI (1969) in the Substantia gelatinosa of the spinal cord. The presence of axo-axonic contacts in trigeminal nuclei has recently been confirmed by GOBEL and DUBNER (1968, 1969), who added the observation that axo-axonic contacts may not only be found on primary afferents but also on corticofugal fibres producing PAD of trigeminal afferents.

The existence of axo-axonic contacts at the relay nuclei of primary afferents of other cranial nerves has not been reported so far. But it has been known for some time that axo-axonic contacts exist in the lateral geniculate body (SZENTÁGOTHAI, 1962; COLONNIER and GUILLERY, 1964). The synaptic endings of the optic afferents are of complex glomerular structure (Szentágothai, 1963; Peters and Palay, 1966) and numerous axo-axonic contacts are found on the optic nerve terminals in the glomeruli. Presynaptic depolarization of optic nerve terminals in the lateral geniculate body has frequently been reported (see chapter 3.7), and it might be assumed, that this depolarization was due to an activation of the axo-axonic contacts of the glomerular complexes, but it has been pointed out that by structural standards the optic nerve terminal mostly appeared to be presynaptic (Szentágothai et al., 1966; Szentágothai, 1968; Hámori, 1968).

In the ventro-basal complex of the thalamus, electrophysiological evidence has been presented that presynaptic inhibition is exerted onto the terminals of lemniscal fibres, presumably by a pathway leading from lemniscal collaterals to a local interneurone, which in turn makes presynaptic axo-axonic contacts on lemniscal terminals (Andersen, Eccles and Sears, 1964, see chapter 3.6). Tömböl (1967) has demonstrated the existence of the histological structures required by this hypothesis, but it remains to be shown that the axon terminals that are postsynaptic in the axo-axonic contacts are lemniscal terminals. At various other sites of the thalamus, axo-axonic contacts having specific structural differentiations have been found (Majorossy et al., 1965; Pappas et al., 1965; Tömböl, 1967). They are particularly frequent in the complex synaptic groupings first described in the pulvinar and designated synaptic glomeruli by Majorossy et al. (1965). So far, no experimental evidence is available which could shed any light on the functional role of these axo-axonic structures.

2.3 Properties of the Inhibitory Axo-Axonic Synapses

2.3.1 Location of Primary Afferent Depolarization

The exact localization of the site of maximum depolarization produced in primary afferent fibres has been elucidated by a systematic investigation of the intraspinal potential fields which were generated either by muscle or by cutaneous afferent volleys, and by testing the excitability changes of afferent fibres at various points along their intraspinal course using the technique developed by Wall (1958). The results gain their significance from the concurrent observations reported later in this review that, when appropriate afferent volleys are used, the depolarization is exerted nearly exclusively either on Ia or Ib fibres from muscle or on Group II fibres from cutaneous nerves. These three fibre types end at different sites in the spinal cord and, consequently, a depolarization of their terminals will produce different intraspinal potential fields.

The intraspinal potential fields produced by Group I afferent volleys from muscle were analyzed by Eccles, Magni and Willis (1962). They found that after the subsidence of the various postsynaptic potentials a potential field can be recorded in the spinal cord (Fig. 10A, B) which is precisely explicable by the postulate that it is generated by a system of core conductors oriented in a dorsomedial-ventrolateral direction and depolarized at or near their ventral ends, as illustrated diagrammatically in Fig. 10C. By subsequent testing of the excitability changes of the Group I afferent fibres from muscle they showed that the core conductors were indeed Group I fibres and that the observed potential field was produced by depolarization of their ventral segments, and not by a hyperpolarization of their dorsal ones. More detailed tests of the

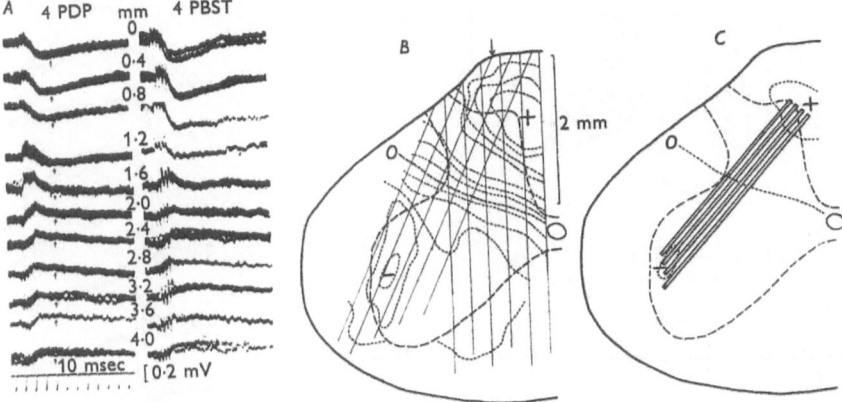

Fig. 10A–C. Intraspinal field potentials produced by Group I afferent volleys from muscle. A Specimen records of the field potentials generated by four PDP and by four PBST Group I volleys (280/s) at the indicated depths along the track marked by arrow in B. Records were made at each 0.2 mm, only every second record being shown in A. Upward deflexions signal negativity relative to the indifferent earth lead. B Contour diagram with equally spaced iso-potential lines for the potential field that is produced by four PDP volleys and recorded as in A for a series of tracks as shown, six being perpendicular and five oblique. The potentials are measured at the arrows in A, 23 ms after the last volley so as to avoid distortion by focal synaptic potentials. C Diagram showing location of dipoles that would give the potential fields of B (ECCLES, MAGNI and WILLIS, 1962)

excitability changes of a population of Ia fibres (Fig. 11) revealed that the principal action of the depolarizing mechanism is exerted on the ramifications of the Ia fibres in the motoneuronal nucleus, but that a second site may be located in the intermediate nucleus where Ia fibres end on interneurones (ECCLES, SCHMIDT and WILLIS, 1963 b).

The excitability of Group Ia fibres could be tested without interference from other muscle afferent types by inserting the stimulating electrode into the motornucleus. There is no comparable location in the spinal cord where Ib fibres may be found in isolation from Ia fibres. However, the excitability changes occurring in Ib fibres can be studied when the antidromic volley produced by central stimulation is "filtered" by an orthodromic Ia volley, which collides with and extinguishes the antidromic Ia spikes before they reach the recording electrodes. The application of this technique together with studies of field potential has shown that the depolarizing synapses on the Ib fibres were probably located in the intermediate nucleus. The increase in excitability was largest here and the negative focus of the potential field was also in this zone (Eccles, Schmidt and Willis, 1963a).

Systematic investigations of the intraspinal potentials produced during the primary afferent depolarization of Group II cutaneous afferent fibres have shown that the maximum zones of positivity and negativity lie fairly close on each side of a zero potential line that runs obliquely across the spinal cord,

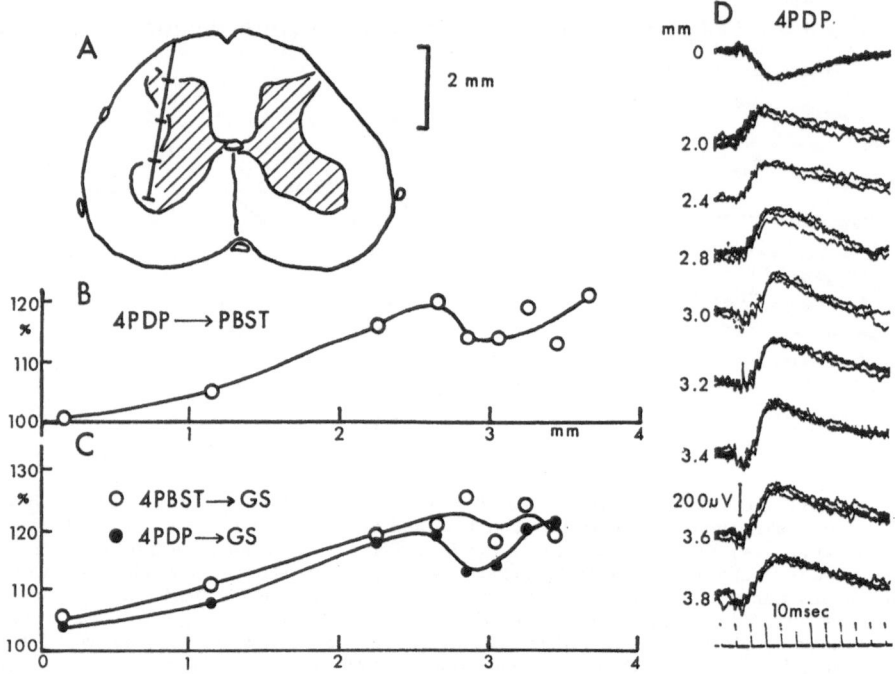

Fig. 11A–D. Locations of depolarizing foci on muscle primary afferents. B and C are excitability increases plotted against the depth in mm along a microelectrode track somewhat more medial than that in A, where the depth in mm is marked. The conditioning stimuli and the fibre terminals on which they were tested are indicated for each series. In D the series of focal potential records with three or four superimposed traces were at the indicated depths, in mm, along the track shown in A and were evoked by four PDP Group I volleys at 300/s. A is a traced enlargement of the section through the spinal cord with the microelectrode *in situ*. Upward deflections in D signal negativity. Same potential and time scales for all traces (Eccles, Schmidt and Willis, 1963b)

being more ventral medially than laterally (Fig. 12, Eccles, Kostyuk and Schmidt, 1962a). This reversal line lies at a depth of about 1 mm below the cord dorsum and about 1 mm more dorsal than the reversal line observed for the field potential, which is generated by Group I afferent volleys from flexor muscles (Fig. 10). The direction and location of the potential fields generated by cutaneous volleys agree closely with the direction and termination of the collaterals of the cutaneous afferent fibres that enter the dorsal horn after branching from the parent fibres in the dorsal column. Excitability testing (Wall, 1958; Eccles, Schmidt and Willis, 1963e) also showed that the maximum depolarizing action of cutaneous afferents occurs at a depth of

1.2–1.7 mm from the cord dorsum, which accords well with the potential fields. If dorsal roots were stimulated instead of peripheral nerves (HOWLAND et al., 1955; WALL, 1962), the fields produced by single shocks resembled those obtained upon single cutaneous nerve volleys since under these conditions the PAD produced by the Group II cutaneous afferents is much more prominent than that produced by muscle afferents.

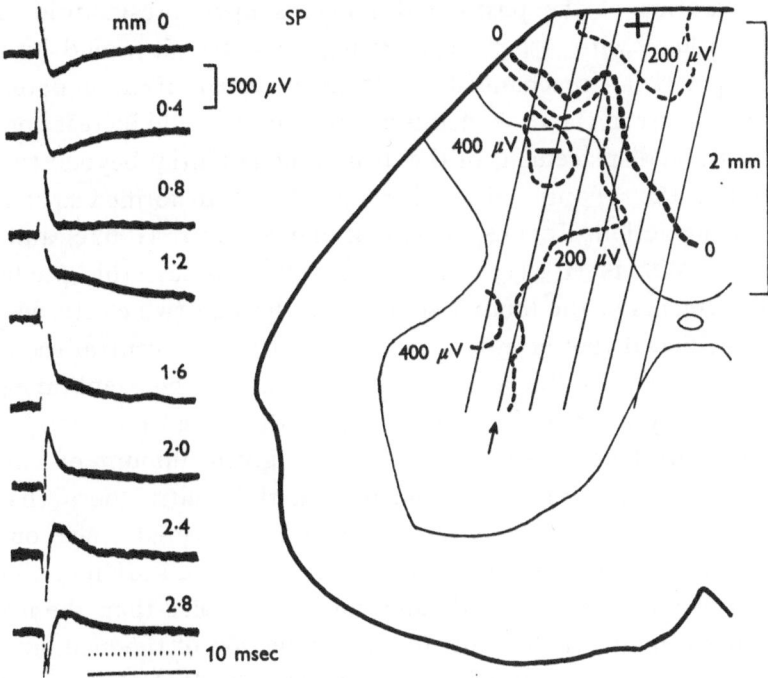

Fig. 12. Field potentials produced by single afferent volleys from a cutaneous nerve. The left-hand column shows specimen records of the field potentials generated by a stimulus to the SP nerve of four times threshold strength at the indicated depths along the track marked by an arrow in the contour diagram. Records were made at each 0.2 mm, but only every second record is displayed. Upward deflections signal negativity relative to the indifferent earth lead. The contour diagram is derived from measurements of a series of six tracks at 0.25 mm intervals, as shown by the oblique lines. The potentials were measured 40 ms after the volley in order to avoid distortion by focal synaptic potentials, but as a consequence the positive potentials in the cord dorsum were underestimated. After the last (most medial) track the microelectrode was left in the spinal cord and its position was determined in the histological preparation
(ECCLES, KOSTYUK and SCHMIDT, 1962a)

Finally, studies of potential fields and excitability measurements in the cuneate nucleus should be mentioned because the histological structure of this nucleus gives particularly favourable conditions for identifying the location of the depolarization exerted onto the cutaneous afferent terminals ending in this nucleus (ANDERSEN et al., 1964a, b). It was shown that the results obtained in these studies could all be explained by the postulate that the depolarization of the cutaneous afferent fibres, induced either by peripheral or cortical stimulation, occurred at their presynaptic terminals ending on cuneate neurones. The importance of these findings relative to electronmicroscopic observations on the ultrastructure of synapses in the cuneate nucleus (Walberg, 1965) has been outlined in chapter 2.2.2 and the functional organization will be dealt with in chapter 3.3.

2.3.2 The Nature of the Presynaptic Depolarizing Potential

This section summarizes experiments which were designed to explore the mode of operation of the postulated synapses upon presynaptic terminals. First, on analogy with other depolarizing synapses, it would be expected that the depolarizing mechanism would have an equilibrium potential. An increase or decrease of the fibre membrane potential should increase or decrease the PAD, and on displacement of the membrane potential beyond the equilibrium potential, the synaptic mechanism should be transformed so as to have a hyperpolarizing action. Eccles, Kostyuk and Schmidt (1962c) and Eccles, Schmidt and Willis (1963c) made an attempt to test this prediction by polarizing segments of the lumbar spinal cord through two electrodes situated medial to the dorsal root entrance and ventral from the ventral roots, respectively. It was shown that the applied currents changed the membrane potential of muscle primary afferent fibres not only at the dorsal root entry zone but also in the central horn region, but for any given amount of current the membrane potential changes were smaller in the ventral horn than in the dorsal parts of the fibres. (This result had been expected since, on account of its large surface area, the terminal ramification of an afferent fibre in the ventral horn would offer a considerably lower resistance than the myelinated dorsally lying segment of the same fibre). Interaction of the membrane potential changes induced by polarizing current with the PAD failed to show the expected increase of the DRP when the terminal membrane potential was hyperpolarized, nor did they reverse the DRP when the membrane potential was depolarized, though it was relatively easy to produce a large decrease in the DRP and even to reduce it to zero. This result may mean merely that the polarizing current blocked the intramedullary transmission of the impulses that generate the DRP. No explanation can be offered for the failure to increase the DRP, but the effect may be comparable to that observed with motoneurones where hyperpolarization usually failed to cause an increase in the EPSP though the membrane potential was displaced further from the equilibrium potential for the EPSP (Eccles, 1957; 1961b; Curtis and Eccles, 1960). Possibly the polarizing current also may cause depression of synaptic transmission along the presynaptic inhibitory pathway. Comparable and equally inconclusive results were obtained in experiments where the influence of polarizing currents onto the DRP of the isolated hemisected spinal cord of the toad was studied (Schmidt, unpublished).

These experiments fail to provide strong evidence that the presynaptic depolarization is due to an ionic mechanism with an equilibrium potential.

Therefore, a more recent attempt of ZIMMERMANN (1968a) to clarify this problem should be mentioned. The excitability changes of single afferent fibres due to PAD were measured with and without a concurrent polarizing current applied through the testing extracellular electrode (Fig. 13). Under these conditions, the PAD seemed to reverse its direction during depolarization of the afferent fibre, and the author came to the conclusion that the equilibrium potential for the PAD is somewhere between $+2$ and $+10$ mV from the resting potential, probably around $+6$ mV.

During after-hyperpolarization induced by tetanic activation of dorsal root fibres, the PAD is often increased to more than twice its control size (ECCLES

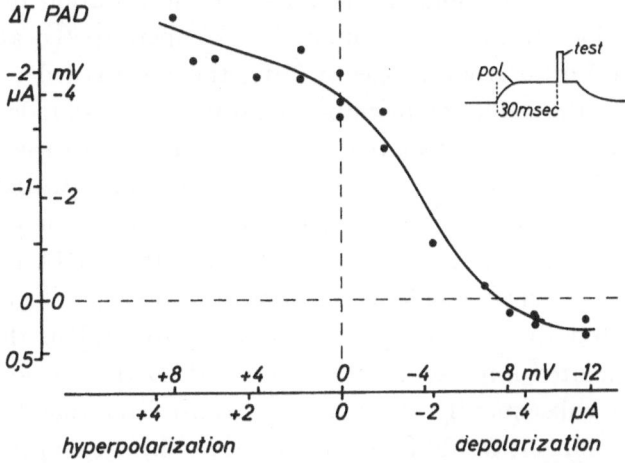

Fig. 13. Modification of the PAD-induced excitability changes of a single primary afferent fibre by polarizing currents. The excitability changes (ΔT, left-hand scale of the ordinate) were tested 30 ms after a single stimulus to the tractus gracilis (cf. Fig. 5). As indicated by the inset polarizing currents were given through the testing microelectrode. The amount and polarity of the polarizing currents is given by the lower scale of the abscissa. The upper scale of the abscissa and the right-hand scale of the ordinate give the approximate membrane potential changes induced by the PAD in the polarizing currents. The tip of the testing microelectrode was at a depth of 1.6 mm from the cord dorsum. The fibre came from a hair follicle receptor and had a peripheral conduction velocity of 57 m/s (ZIMMERMANN, 1968a)

and KRNJEVIĆ, 1959a; ECCLES, MAGNI and WILLIS, 1962; ECCLES, SCHMIDT and WILLIS, 1963c), and it was already suggested by ECCLES and KRNJEVIĆ (1959a) that the increase in the PAD is sufficiently explained by the after-hyperpolarization. Hence it might be inferred that the after-hyperpolarization had more than doubled the separation of the membrane potential from the equilibrium potential for the ionic mechanism producing the PAD. However, the rising slope of the PAD was never increased as much as the height of the PAD, and was sometimes even decreased. The large increase in height was mainly due to the longer duration of the rising phase which was invariably observed. It has to be concluded, therefore, that the increase in membrane potential brought about by the after-hyperpolarization caused no more than a small increase in the currents generated by the presynaptic depolarizing synapse. Excitability testing showed that the after-hyperpolarization of the nerve terminals is larger than in the main shaft of the fibres (Wall and Johnson, 1958); nevertheless it would not be expected to be in excess of 10 mV, for with intrafibre recording it was seldom more than 5 mV (Eccles and Krnjević, 1959a, b). From these results it would appear that the equilibrium potential for the PAD would be at a depolarization of at least 30 mV relative to the resting membrane potential.

An action potential propagating down an afferent fibre to its central terminal might be expected to erase all electrotonic potentials in the fibre. Its action on the PAD provided a further clue to the nature of the mechanism producing the depolarization, particularly on the duration of the presumed transmitter action. If an action potential was superimposed at various times during the PAD of a Group Ia nerve fibre, the PAD failed to recover to its control level but the reduction in amplitude was very small indeed, particularly when the impulse was interpolated early with respect to the PAD (Eccles, Schmidt and Willis, 1963c; Schmidt, Senges and Zimmermann, 1967a). Somewhat comparable observations have been reported for the action of an interpolated afferent volley in partly destroying the DRPs recorded from a frog dorsal root (Eccles and Malcolm, 1946). Eccles, Schmidt and Willis (1963c) suggested that the simplest hypothesis to explain these findings is that the interpolated impulse destroys all the PAD that is preformed in that fibre, but that subsequently the lingering transmitter rebuilds much of the depolarization. Thus it would be envisaged that the transmitter continues to act throughout the whole duration of the PAD. This mechanism ensures that repetitive afferent impulses appearing at the fibre terminals during PAD are all well inhibited.

Finally, experiments should be mentioned which indicate that presynaptic terminals can also be influenced by postsynaptic neurones. Antidromic volleys in ventral roots of the cat spinal cord do not induce DRP or DRRs. However, if the primary afferent fibres are depolarized by a conditioning volley in a neighbouring dorsal root or rootlet, antidromic stimulation of ventral roots induces DRRs in the depolarized fibres (Decima, 1969; Decima and Goldberg, 1969, 1970). The latency of the DRRs induced by motoneurone excitation is extremely short, and it was proposed, therefore, that the effects were due to some kind of electrical coupling between motoneurones and presynaptic terminals.

In this section experiments have been reviewed which were designed to elucidate the nature of the depolarizing potential induced in primary afferents. The results have been described in some detail, to show that they are in accordance with the hypothesis that this potential can be attributed to the prolonged action of a chemical transmitter substance, which operates in a manner comparable with other depolarizing transmitters, namely by effecting a high permeability to ions; but it has to be appreciated that the mechanism generating the PAD is far from understood. In view of the functional significance attributed to presynaptic inhibition, the unsatisfactory knowledge of the PAD generating mechanism remains a challenge for all neurophysiologists interested in this field.

2.3.3 Spike Size and Transmitter Release

During PAD a reduced amount of transmitter is liberated when an impulse propagates down to a depolarized presynaptic nerve terminal. In the extreme situation, the impulse is blocked before reaching some or all of the terminals (BARRON and MATTHEWS, 1935; BROOKS, ECCLES and MALCOLM, 1948; HOW- LAND et al., 1955), but in most instances the terminals are invaded by a spike potential of decreased size, which releases a reduced amount of transmitter. There is much indirect evidence in the literature that such a positive correlation exists between presynaptic spike size and synaptic efficacy (LLOYD, 1949; ECCLES and RALL, 1951; DEL CASTILLO and KATZ, 1954a; LILEY, 1956; ECCLES and KRNJEVIĆ, 1959a, b; BENOIT and MAMBRINI, 1970). Even more relevant is the direct evidence that has been produced by taking advantage of the very large size of both the presynaptic and postsynaptic components of the giant synapses of the squid stellate ganglion (HAGIWARA and TASAKI, 1958; TAKEUCHI and TAKEUCHI, 1962; BLOEDEL et al., 1966, 1967; LLINÁS, 1968; KATZ and MILEDI, 1966, 1967c). Simultaneous intracellular recording from the pre- and postsynaptic side revealed that changes in the presynaptic spike potential induced profound changes in the size of the excitatory post-synaptic potential. Qualitatively similar results were obtained at the neuromuscular junctions of rat, frog and crayfish (HUBBARD and SCHMIDT, 1963; DUDEL, 1965; BRAUN and SCHMIDT, 1966; KATZ and MILEDI, 1967a, b; HUBBARD and WILLIS, 1968).

The action potential amplitude of dorsal root fibres depends on the resting membrane potential: hyperpolarization of the resting membrane potential results in an increased spike amplitude, whereas depolarization gives a reduced spike amplitude (ECCLES and KRNJEVIĆ, 1959b; ECCLES, KOSTYUK and SCHMIDT, 1962c). Intrafibre recording revealed that during PAD the spike potential was reduced by an amount approximately equivalent to the de- polarization (ECCLES, SCHMIDT and WILLIS, 1963b). If the PAD was mimicked by an electrical polarization of the spinal cord, the changes of the spike ampli- tude of dorsal root fibres were accompanied by relatively large variations in the size of their excitatory synaptic actions (Fig. 14). If it is assumed that the spike potential of the afferent fibre terminals varied with the membrane poten- tial, in the same manner as occurred for their main shafts in the dorsum of the cord, the results illustrated in Fig. 14 provide the first direct evidence in support of the hypotheses that attribute increases or decreases of the synaptic excitatory actions on motoneurones to changes in the presynaptic spike potential. [When the polarizing currents were applied to dorsal roots (Eccles and Krnjević, 1959b) there was little or no change in the EPSP of motoneurones, presumably because the electrotonic decrement along the Ia afferent fibres to their terminals in the motoneurone nucleus was too great].

The reduction of the EPSP during strong presynaptic inhibition (Eccles, Eccles and Magni, 1961) was always larger than that found in the polarizing

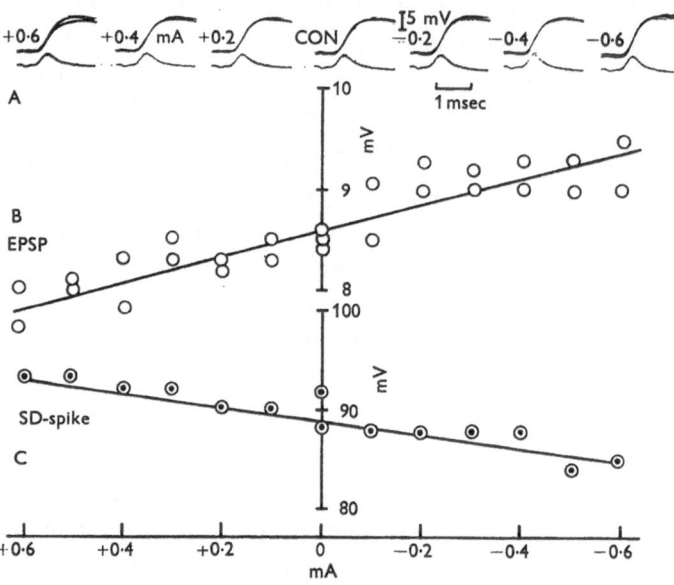

Fig. 14 A–C. Changes produced in monosynaptic EPSP by polarizing current across the cord. The polarizing current was applied through two electrodes, one situated medial to the dorsal root entrance, the other just lateral from the ventral roots. Intracellular recording from a motoneurone, the membrane potential being −70 mV. A shows specimen records of EPSPs under the influence of increasing currents in both directions, as indicated (in mA) for each record, CON being the control value. Each record consists of many superimposed faint traces. The upper traces are the intracellular records, which are differentiated in the lower traces. In B the amplitudes of the EPSPs (in mV) for the series partly shown in A are plotted against the direction and strength of the polarizing currents. C shows the amplitudes of antidromically evoked SD-spikes measured under the influence of the same currents. In A, B and C (+) and (−) indicate the polarity of the dorsal electrode. Voltage calibration in A is for intracellular recording only (Eccles, Kostyuk and Schmidt, 1962c)

experiments. There are several possible explanations of this difference. In the first place, in presynaptic inhibition the depolarization of the presynaptic terminals may be larger. Secondly, since with presynaptic inhibition the depolarization of the afferent fibres is actively produced by a transmitter substance, the presynaptic spike would be decreased not only by the diminution of the membrane potential (as occurs with the polarizing current), but also as a consequence of the increased ionic conductance, just as occurs with the muscle impulse at an activated motor end-plane (Fatt and Katz, 1951; del Castillo and Katz, 1954b) and also with synaptic action on nerve cells (Fadiga and Brookhart, 1960; Nishi and Koketsu, 1960). In a manner similar to that at the presynaptic inhibitory synapse of the crayfish (Dudel,

1963, 1965), the conductance increase may be more important than the membrane potential change. Furthermore, during strong PAD, collision with an antidromic DRR or block of conduction may prevent altogether the invasion of some or all of the synaptic terminals of primary afferent fibres, a possibility firmly advocated by WALL and his colleagues (HOWLAND et al., 1955; WALL et al., 1955; WALL, 1964). Finally it should be pointed out that the liberation of transmitter may have a very steep relationship to the size of the spike potential in the synaptic terminals, even a fourth power relationship (LILEY, 1956); spike augmentations of 10–25 % have been shown to enlarge the transmitter output at the neuromuscular junction by 200–300 % (HUBBARD and SCHMIDT, 1962, 1963). KATZ (1962) has suggested that, as a consequence, the relatively small depolarization produced in presynaptic inhibition may nevertheless have a large depressant action on transmitter liberation and so on the EPSP.

2.4 Central Pathways Responsible for Primary Afferent Depolarization
2.4.1 The Mode of Operation of the Presynaptic Pathway

The neuronal pathways for primary afferent depolarization have properties corresponding to those characterizing the polysynaptic pathways of the flexor reflex. First of all, there is a central latency of several milliseconds. The shortest central latencies were found in the cuneate nucleus, where 2.0–2.2 ms elapsed between the arrival time of the fastest component of the afferent volley to the cuneate nucleus, and the onset of PAD recorded intracellularly from a cuneate tract fibre (Fig. 4, ANDERSEN, ECCLES, SCHMIDT and YOKOTA, 1964b). At the lumbar segmental level the central latency of the PAD ranged from 2.0 to 3.0 ms for cutaneous primary afferent fibres (KOKETSU, 1956; ECCLES and KRNJEVIĆ, 1959a), and from 4.0 to 5.0 ms for muscle primary afferents (ECCLES, MAGNI and WILLIS, 1962). Furthermore, there are other features of polysynaptic pathways such as temporal and spatial facilitation, posttetanic potentiation, and depression during repetitive stimulation. Temporal and spatial facilitation can be more easily demonstrated in those pathways leading from Group I muscle afferents to Group I afferents, than in those from cutaneous to cutaneous afferent fibres, mainly because the latter reflex pathways require only very few impulses for maximal activation (SCHMIDT, SENGES and ZIMMERMANN, 1967b; JÄNIG et al., 1968b).

The effectiveness and time course of the temporal facilitation of DRPs evoked by electrical stimulation of Group I muscle afferents is illustrated in Fig. 15 (ECCLES, SCHMIDT and WILLIS, 1963b). Similar results were obtained by ECCLES, MAGNI and WILLIS (1962) using other methods for demonstrating the PAD, including intrafibre recording from Group Ia primary afferent fibres. In all these experiments, 4–6 afferent volleys at 200–300 Hz were sufficient to attain nearly maximum potentiations, and the facilitation was usually most prominent when the effect of a single volley was small. A striking example for spatial facilitation of presynaptic inhibition is shown in Fig. 16, where the monosynaptically evoked EPSP in a plantaris motoneurone was subjected to presynaptic inhibition from ST and PDP afferents. Post-tetanic potentiation of PAD, after prolonged tetanization of the conditioning pathway, has been demonstrated in several investigations (Eccles, Eccles and Magni, 1961; Eccles, Magni and Willis, 1962; Eccles, Schmidt and Willis, 1963b), and

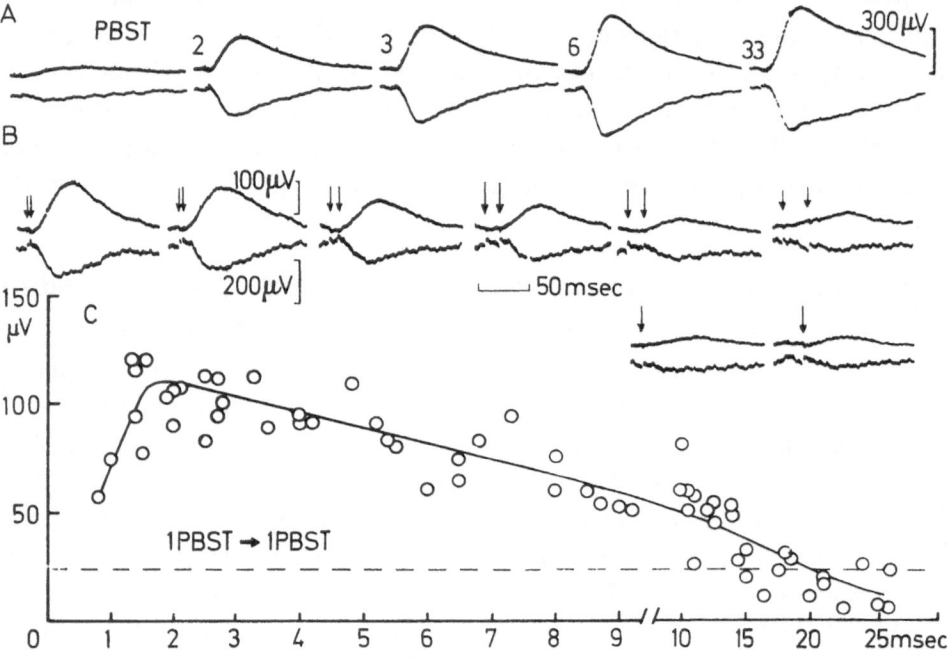

Fig. 15A–C. Temporal facilitation of PAD. A DRP (upward deflections) and P-waves produced by the indicated number of maximum Group I volleys in PBST nerve at 300/s. Voltage scale is 300 µV for the DRP and 500 µV for the P-waves. B and C Two Group I volleys were set up at the intervals indicated by the arrows in the specimen records of B and the evoked DRP and P-waves were recorded as in A. The control responses to each volley alone are shown with the records below single arrow). In B the sizes of the DRP produced by the second volley were determined by subtraction of their initial control response from the combined response and plotted against the stimulus intervals. The broken line shows the size of the DRP produced by the second volley alone. It is seen that at about 20 ms facilitation gives place to depression (Eccles, Schmidt and Willis, 1963b)

the same papers also give examples for depression of the PAD generating pathways during repetitive stimulation (see also Decandia et al., 1968). Thus, in every respect, the pathways leading from primary afferents to the PAD generating mechanism behave as other polysynaptic reflex pathways of the central nervous system. The central latencies of the PAD are not shorter than 2.0 ms, which gives time for pathways that include two synapses in serial order, as has been postulated for the simplest pathways for PAD production in the spinal cord and the cuneate nucleus (Eccles, Kostyuk and Schmidt, 1962a, Andersen, Eccles, Schmidt and Yokota, 1964b).

So far, very little is known about the pathways leading from cutaneous C-afferents to myelinated afferents from either skin or muscle nerves. It has only been demonstrated that the PAD evoked by C-fibre activity converges and interacts with the PAD evoked by cutaneous A-volleys in myelinated afferents (JÄNIG and ZIMMERMANN, 1971).

Evidence has been presented that in the unanaesthetized spinal animal the pathways leading to Ia terminals in the motoneuronal nuclei exhibit considerable spontaneous activity. This activity causes excitability fluctuations of Ia fibre terminals, and it is mainly responsible for the variability of the

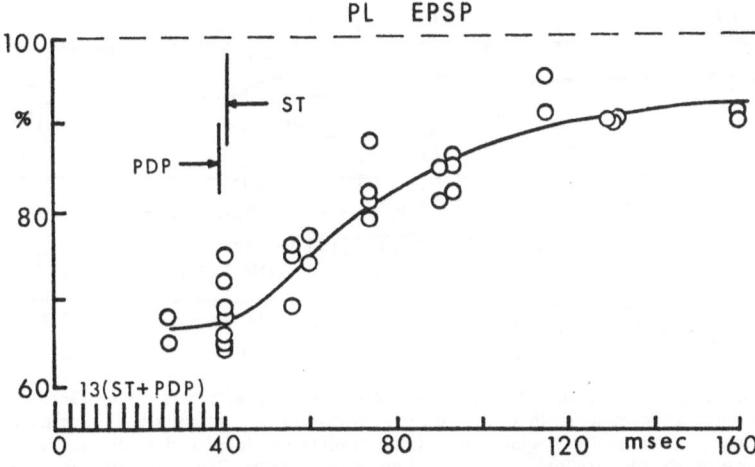

Fig. 16. Spatial summation of presynaptic inhibition of monosynaptic EPSPs. The plotted points give the per cent depression of the monosynaptic EPSPs in a plantaris motoneurone at various intervals after conditioning by 13 Group I volleys in the semitendinosus and peroneal flexor nerves (ST + PDP), as indicated on the base line. The arrows labelled PDP and ST each give the mean values of six separate measurements of the EPSP depressions by 13 PDP and 13 ST volleys respectively at a test interval of 40 ms, the length of the vertical lines indicating the respective ranges of the measurement (ECCLES, SCHMIDT and WILLIS, 1963b)

monosynaptic reflex responses observed under these conditions (RUDOMIN and DUTTON, 1967, 1968, 1969a, b). Conditioning afferent volleys in muscle and cutaneous nerves reduce both the reflex variability and the excitability fluctuations. These effects as well as the concurrent changes caused in the correlation between monosynaptic reflexes can be fully explained by the inhibitory influences which these conditioning volleys exert onto the interneurones affecting transmission from Ia fibres to motoneurones via PAD (RUDOMIN and DUTTON, 1969a, b; RUDOMIN et al., 1969). Corresponding findings have now been described in the somatosensory system (ROWE, 1970).

2.4.2 Interneurones on the Pathway of PAD

In the studies on the functional organization of primary afferent depolarization a very specific pattern has emerged for the rôle of the various afferent fibre groups in producing and receiving this depolarization (see section 3 of this review). The principal findings are that Group I a and I b muscle afferents are depolarized mainly from Group I muscle impulses (Figs. 24, 26), while the cutaneous afferents and possibly the other fibres of the FRA-system (ECCLES and LUNDBERG, 1959, cf. 3.1.3) are mainly depolarized by the FRA-system but also by Group I b muscle impulses (Fig. 27). Since the central latency of PAD is usually 2.0–3.0 ms, there should be at least two interneurones in serial order on the central pathway. Obviously the last neurones of the PAD reflex chains, which make synaptic contacts with the afferent fibre terminals, have

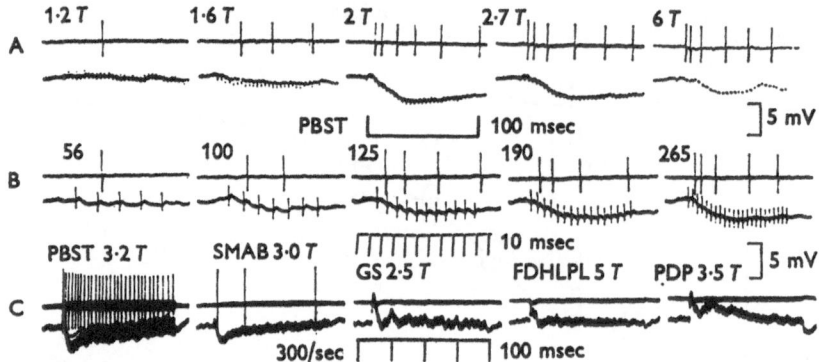

Fig. 17A–C. Discharge patterns of a D-type interneurone. The upper traces are extracellular recordings from the D interneurone at a depth of 2.15 mm. The lower traces are the cord dorsum potentials at L_7 segmental level. A shows the response to tetanic stimulation (265/s) of PBST nerve with stimulus strengths given relative to threshold. In B the stimulus strength to PBST was kept constant at 2.15 T, just maximal for Group I, and various tetanus frequencies were employed: numbers show frequencies in c/s. C shows the response to tetanic stimulation (300/s) of different muscle nerves at much slower sweep speed. The nerve and the stimulus strength relative to threshold are indicated on each record. Upper voltage calibration is for upper trace in A; lower voltage calibration is for upper traces in B and C
(ECCLES, KOSTYUK and SCHMIDT, 1962a)

to have excitatory inputs which are in agreement with the functional organization outlined above.

ECCLES, KOSTYUK and SCHMIDT (1962a) carried out a search for such interneurones. Those cells that could be classified as belonging to one or the other reflex chain of PAD were labelled D-cells. Fig. 17 shows records from a D-cell that was listed as a possible candidate to occupy a second or later place of the serial order in the polysynaptic PAD pathway of Group I muscle afferents because it fulfilled the criteria expected for such an interneurone. In the same series of experiments other D-cells were identified which fulfilled the criteria for being candidates on the second or later places of the pathways of the cutaneous and high threshold muscles (the flexor reflex, FRA) afferents. The majority of both types of cells were located in the base of the dorsal horn at a depth of 1.65–2.5 mm from the dorsal surface.

The evidence presented in the two preceding chapters is summarized in Fig. 18 which on the left-hand side gives diagrammatically the postulated

segmental interneuronal pathways concerned in the depolarization of Group I
afferent fibres from muscle, and, on the right-hand side, those concerned in
the depolarization of cutaneous fibres by volleys in cutaneous fibres and
Group II afferent fibres from muscle. It has to be appreciated that most
probably there exist much more complex paths of activation of the interneurones making axo-axonic contacts on myelinated afferents than the simplest
possible pathways in the diagram. The afferent C-fibres from skin, and their

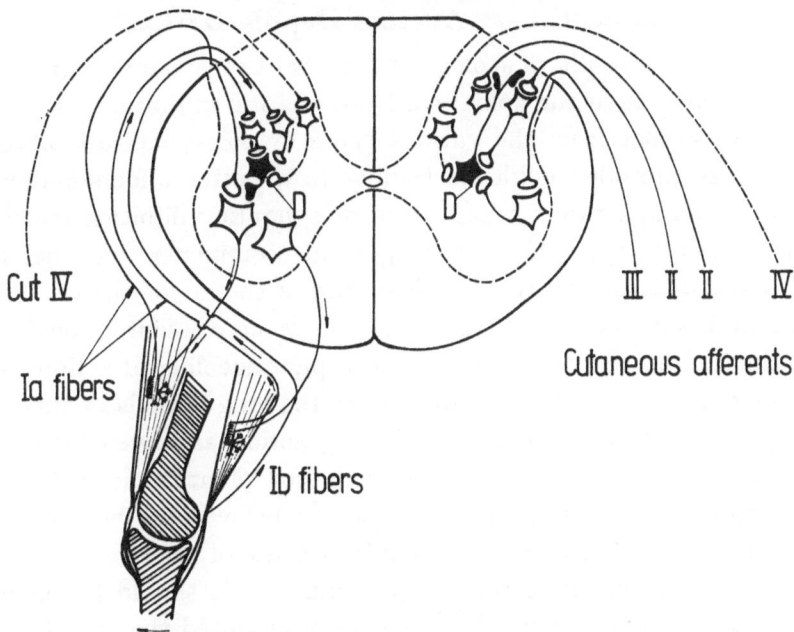

Fig. 18. Schematic diagram illustrating interneuronal pathways of PAD. The left-hand side illustrates
the convergence of flexor I a and I b fibres onto a D type interneurone which in turn makes axo-axonic
contacts on the terminals of an afferent I a fibre from extensor muscle. The pathway labelled Cut IV
indicates that cutaneous C afferents may also converge on Group I afferents from muscle. On the right-
hand side pathways for presynaptic inhibitory actions on cutaneous primary afferents are shown. Further
discussion in the text. (Modified from ECCLES, KOSTYUK and SCHMIDT, 1962a)

central connections are drawn as dashed lines in order to indicate the lack
of information on the length of the interneuronal chains used by these inputs.

WALL (1962, 1964) maintains that the activity of the small cells of the
Substantia gelatinosa (lamina II, SG-cells) controls the membrane potential
of the cutaneous afferent terminals and that these cells are in turn excited
monosynaptically by large afferents. Since it has not been possible, so far,
to record from the small SG-cells, only indirect evidence has been presented
to support these hypotheses. The findings described above and summarized
on the right hand side of Fig. 18 do not exclude the possibility that SG-cells
are interconnected between D-cells and primary afferent terminals, but the
more attractive suggestion appears to be that the axons of the SG-cells make

excitatory synaptic contacts with the dendrites of those lamina IV cells which send their axons to the synaptic complexes in lamina II to establish axo-axonic contacts with primary afferents (Fig. 7, RÉTHELYI and SZENTÁGOTHAI, 1969). The D-cells, i.e. the pyramidal cells P in Fig. 7, would thus act as the final common path for segmental as well as suprasegmental (DES in Fig. 7) influences resulting in PAD of myelinated cutaneous afferents.

2.5 Pharmacology of Presynaptic Inhibition and Primary Afferent Depolarization

The studies on the pharmacological aspects of presynaptic inhibition have not only shown that in the spinal cord presynaptic inhibition has a pharmacology quite distinct from the various types of postsynaptic inhibition, but they have also provided evidence leading to a better understanding of the organization and synaptic mechanism of presynaptic inhibition (cf. SCHMIDT, 1964, 1965 a; ECCLES, 1965). Two general comments on the methods used in many of these pharmacological studies should be made: firstly, the recording of DRPs has proved to be the most convenient, though it is an indiscriminative procedure for the study of pharmacological actions on PAD. These potentials are sufficiently stable over the long periods of time that are required for the full development of a drug action and even for its decline. They have therefore been used extensively for pharmacological studies on PAD. Secondly, many of the results reported below have been obtained on the amphibian spinal cord *in vitro*, mainly because of the obvious advantages of such a preparation in pharmacological studies. It should be pointed out, however, that, although it has been generally accepted that the PAD in the feline and amphibian spinal cord are both generated by a similar mechanism and have a similar physiological significance, so far only indirect evidence has been brought forward relating the amphibian PAD to a presynaptic inhibitory mechanism (cf. HOLEMANS and MEIJ, 1968a, b).

2.5.1 Anaesthetics and Other Depressant Drugs

ECCLES and MALCOLM (1946) reported that pentobarbitone (Nembutal) greatly prolonged the decay of the DRP. This observation has meanwhile been confirmed both in the cat (LLOYD, 1952; ECCLES, SCHMIDT and WILLIS, 1963 f) and in the isolated frog and toad spinal cord (SCHMIDT, 1963, 1964; GRINNELL, 1966; RICHENS, 1969). Small doses of Nembutal not only prolonged but also increased the PAD, whereas higher doses led to a further prolongation together with a reduced amplitude. Other barbiturates and chloralose had similar effects. The amplitude and time courses of presynaptic inhibition followed closely the variations of the DRP (ECCLES, SCHMIDT and WILLIS, 1963 f, Figs. 1–3). At the cuneate nucleus pentobarbital increased the amplitude of the P-wave and prolonged its duration. It also increased and prolonged the concurrent excitability changes of the posterior column tract fibres. Parallel to these events the inhibition of the lemniscal discharge was increased and prolonged (BANNA and JABBUR, 1969).

All other anaesthetics so far tested (chloral-hydrate, paraldehyde, ether, urethane, chloroform, halothane, ethylchloride, methoxyflurane, trichloroethylene, propanidid) shortened the decaying phase of the DRP instead of prolonging it (SCHMIDT, 1963; RICHENS, 1969). This shortening was more obvious the longer the declining phase of the DRP before the application of the anaesthetic. The most likely explanation for this phenomenon is that the anaesthetic curtailed the afterdischarges of the interneurones in the pathways responsible for the PAD. Regularly with smaller doses of anaesthetics the DRPs were also increased. This was probably caused by a hyperpolarization of primary afferent fibres due to the removal of background activity of the interneurones mediating the PAD. Such a removal would increase the efficiency of the primary afferent fibres producing the PAD and would also increase the potential difference between the resting potential and the equilibrium potential of PAD, thus enhancing the synaptic drive during activation. The increase of the PAD under conditions of sedation or light anaesthesia is paralleled by an increase of the presynaptic inhibition of monosynaptic reflexes (MIYAHARA et al., 1966). In addition to some of the anaesthetics already mentioned, these authors also tested the effects of phenobarbital, ethanol, nitrous oxide and magnesium sulphate, all having similar effects.

Depressant drugs had diverse effects on presynaptic inhibition of monosynaptic reflexes. Mephenesin always blocked presynaptic inhibition, even in concentrations that had little effect on unconditioned monosynaptic responses (LLINAS, 1964; RUDOMIN, 1966; MIYAHARA et al., 1966). The latter authors reported that the anticonvulsant drug trimethiadone showed actions similar to those exhibited by the anaesthetics, whereas procaine enhanced presynaptic inhibition in low doses and blocked it in high doses (cf. also GRINNELL, 1966). The tranquillizing and muscle relaxant agent diazepam had particularly pronounced actions on PAD (SCHMIDT, 1965; PIXNER, 1966; SCHMIDT, VOGEL and ZIMMERMANN, 1967; STRATTEN and BARNES, 1968). It was shown that this substance increased considerably lumbar DRPs evoked either by stimulation of muscle or cutaneous nerves or by natural stimulation of skin receptors (Fig. 19). Correspondingly the presynaptic inhibition of monosynaptic reflexes was increased and prolonged. The amount and time course of postsynaptic inhibition of motoneurones was not altered by diazepam, nor were the polysynaptic spinal reflex pathways strongly depressed (NGAI et al., 1966; SCHMIDT, VOGEL and ZIMMERMANN, 1967). It was concluded by the latter authors that the muscle relaxant effect of diazepam was at least partly due to the increase and prolongation of presynaptic inhibition (cf. also HUDSON and WOLPERT, 1970).

There is as yet no explanation for the prolongation of the DRP during barbiturate anaesthesia or following diazepam administration. Eccles, Schmidt and Willis (1963 f) proposed that the prolongation was due either to the inhibition of an enzyme system that destroys the transmitter substance producing the PAD, or that the anaesthetics act by impeding the diffusional spread of the transmitter substance from its site of action (cf. also Schmidt, 1964). Wall (1964) and Mendell and Wall (1964) have postulated that

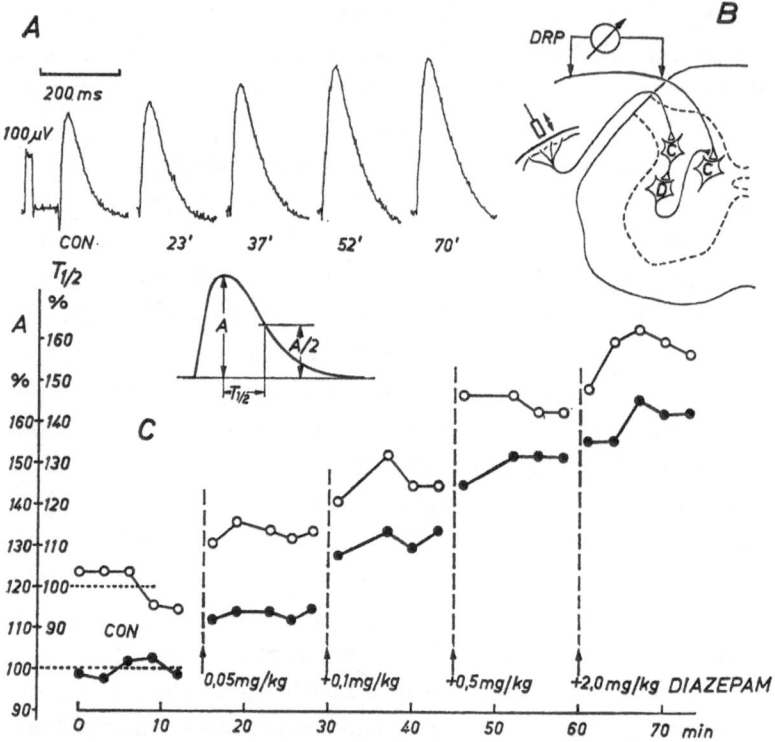

Fig. 19A–C. Action of Diazepam on DRPs. B The DRPs were recorded from a L_7 dorsal rootlet following mechanical stimulation of the central foot pad (700 μ indentation for 5 ms). The specimens in A are $x − y$-plots of averaged DRPs (20 trials each) recorded before (CON) and after 4 consecutive diazepam administrations. The indicated time in min corresponds to that shown in the abscissa of C. C As indicated in the inset the filled circles and the left-hand ordinate plot the amplitude of the DRPs, the open circles and the right-hand ordinate give the time to half decay (Schmidt, Vogel and Zimmermann, 1967)

Nembutal acts by prolonging the discharge of the interneurones on the pathway of PAD. They argue that the declining phase of the PAD is far too long to be a passive process and that it must be sustained by a continuous interneuronal discharge. Eccles, Schmidt and Willis (1963 f) had found that Nembutal invariably decreased the rate of discharge of interneurones firing spontaneously or after orthodromic activation. This fits well with the observation that all anaesthetics, even the barbiturates, lead in the beginning of anaesthesia to a shortening of the PAD together with an increase of its amplitude. Wall (1964) and MENDELL and WALL (1964) also could not find a prolongation of inter-neuronal discharge during Nembutal anaesthesia. They therefore concluded that PAD is not mediated by interneurones like the D-cells proposed by ECCLES, KOSTYUK and SCHMIDT (1962a), but by very small cells like those of the Substantia gelatinosa Rolandi which are too small to be detected by a searching microelectrode. It is difficult to imagine that these cells could be the only ones in the spinal cord which prolong their discharge in the course of barbiturate and chloralose anaesthesia, but behave like all the other cells when exposed to any other type of anaesthetic. Indeed, none of the suggestions proposed so far to explain the prolongation of the PAD following barbiturate appli-cation is particularly convincing. This uncertainty doubtless derives from our ignorance of the neuronal mechanism producing the PAD.

2.5.2 Convulsants

It is well established that strychnine depresses postsynaptic inhibition in the spinal cord, and the convulsant activity of strychnine and similarly acting drugs seems to be fully explained by this depression (for review see CURTIS, 1963; ECCLES, 1964a). There are several early reports in the literature (UMRATH, 1933; BREMER, 1941, 1944, 1953; DUN, 1942; DUN and FENG, 1944; ECCLES and MALCOLM, 1946) of the action of strychnine on the DRP of the frog spinal cord.

All authors agree that the size of DRP evoked by dorsal root stimulation is not appreciably altered, but that the declining phase is greatly prolonged. The P-wave recorded from the cord dorsum is similarly affected (BERNHARD and KOLL, 1953). More recent re-investigations of the action of strychnine on the PAD, both in the cat and in the isolated toad spinal cord, have confirmed the prolongation of the declining phase of the DRP and P-wave, and have shown that the amplitude of the PAD is increased in the early stages of strychnine poisoning, whereas in the later stages there is a reduction of the PAD amplitude (cf. SCHMIDT, 1963, Fig. 7). In the cat spinal cord, the increased and prolonged DRPs are paralleled by an increased and prolonged presynaptic inhibition of monosynaptic reflexes (ECCLES, SCHMIDT and WILLIS, 1963f; Fig. 8). In the cuneate nucleus, the effects of strychnine on PAD and pre-synaptic inhibition resemble those seen at the spinal level (BOYD et al., 1966; BANNA and JABBUR, 1969).

It has been proposed by ECCLES, SCHMIDT and WILLIS (1963f) and SCHMIDT (1963) that the effects of strychnine on PAD and presynaptic inhibition are likely to be caused by a removal of postsynaptic inhibition on the cells of the pathways of PAD, and that strychnine has no direct action on the axo-axonic inhibitory synapses. In the early stages of strychnine poisoning the removal of postsynaptic inhibition will result in an enhanced and prolonged interneuronal discharge and a consequently enhanced and prolonged PAD, whereas in the later stages the PAD will be reduced because of the occlusion produced in the PAD pathways by the considerable spontaneous and convulsive inter-neuronal activity.

Picrotoxin does not affect postsynaptic inhibition, but evidence has been brought forward that this drug interferes with PAD and presynaptic inhibition. In the isolated toad spinal cord, picrotoxin depressed the DRPs in concentra-tions which enhanced the ventral root reflex discharges (Schmidt, 1963; Grinnell, 1966; Tebēcis and Phillis, 1969a). Eccles, Schmidt and Willis

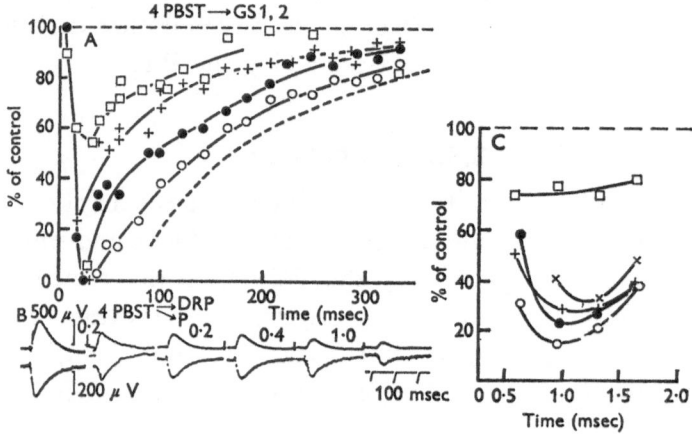

Fig. 20A–C. Action of picrotoxin on the presynaptic and postsynaptic inhibition of monosynaptic reflexes and on DRPs and P-waves. In A the presynaptic inhibitory action of 4 PBST Group I volleys (300/s) was tested by monosynaptic reflexes evoked by 2 GS volleys at 1.5 ms interval and recorded monophasically in the S1 ventral root. The symbols show the relation of the successive i.v. injections of picrotoxin to the inhibitory curves. – – – – control; ○ after picrotoxin 0.2 mg/kg; ● after further 0.6 mg/kg; + after further 1 mg/kg; □ 10 min later. The DRPs and P-waves in B were obtained concurrently with the inhibitory curves in A. C shows the action of successive i.v. injections of picrotoxin on the postsynaptic inhibition, one Q Group I afferent volley inhibiting the monosynaptic reflex produced by a PBST volley. ○ control; ● after picrotoxin 0.08 mg/kg; + after further 0.16 mg/kg; × after further 0.5 mg/kg; □ after strychnine 0.08 mg/kg (Eccles, Schmidt and Willis, 1963f)

(1963f) found in the spinal cord of the spinal cat that picrotoxin reduced the dorsal root potentials and diminished presynaptic inhibition (Fig. 20). With higher concentrations the decreased DRP was accompanied by convulsant activity of the preparation. In the decerebrate cat the depression of segmental DRPs and the removal of presynaptic inhibition of monosynaptic reflexes by picrotoxin appeared even more pronounced than in the spinal animal (Llinás, 1964). At the cuneate nucleus, picrotoxin as well as bemegride and pentylene-tetrazol depressed the P-wave and increased the excitability of cuneate presynaptic terminals produced by conditioning cortical or cutaneous volleys (Banna and Jabbur, 1969, 1970; Banna and Hazbun, 1969). Furthermore, it reduced the inhibition of the lemniscal discharge by conditioning cutaneous sources (Boyd et al., 1966; Banna and Jabbur, 1968, 1969).

These results are all compatible with the conclusion reached by ECCLES, SCHMIDT and WILLIS (1963 f) and SCHMIDT (1963) that picrotoxin produces a genuine depression of the presynaptic inhibitory synapses, and that the convulsant activity of picrotoxin (and similarly acting drugs) is at least in part due to this depression of presynaptic inhibition. It was pointed out that the depression was probably not due to convulsant activity occluding the interneuronal pathways responsible for the PAD, since strychnine enhanced and prolonged the PAD rather than depressing it. Furthermore picrotoxin depressed presynaptic inhibition at a dosage below that causing convulsions. Possibly this action is attributable to a competitive occupation of the receptor sites for the presynaptic inhibitory transmitter substance. Such a mechanism has already been proposed to account for the blocking action of picrotoxin at the inhibitory synapse of the crayfish neuromuscular junction, where GABA is the transmitter (ROBBINS and VAN DER KLOOT, 1958; GRUNDFEST et al., 1959; GRUNDFEST and REUBEN, 1961). A GABA-picrotoxin interaction has recently been demonstrated by GALINDO (1969) in the mammalian central nervous system and by TEBĒCIS and PHILLIS (1969a, b) in the toad spinal cord. And bicuculline, which has been shown to be a relative selective GABA antagonist, and to suppress certain strychnine-insensitive inhibitions in the feline cerebral cortex and cerebellum (CURTIS et al., 1970a, b; see however GODFRAIND et al., 1970), appears to have a powerful action on presynaptic inhibition (Curtis, personal communication). Certainly, the striking difference between the inhibitory depressant actions of strychnine and picrotoxin at least suggests that either different transmitter substances or different receptor sites, or both, are present at postsynaptic and presynaptic inhibitory synapses.

It may well be that the hypotheses outlined above for the action of strychnine and picrotoxin are over-simplifications since several aspects of these hypotheses are questionable. For instance, TEBÉCIS and PHILLIS (1969a) pointed out that in some of their experiments picrotoxin even at high concentrations had no effect on DRPs and they concluded that the action of picrotoxin is more complex than has previously been supposed. Furthermore, as outlined in chapter 1.2, GRANIT and his co-workers have failed to find appreciable amounts of presynaptic inhibition, but instead have described a strychnine-resistant, picrotoxin-sensitive postsynaptic inhibition (KELLERTH, 1965, 1968; KELLERTH and SZUMSKI, 1966a, b). Their conclusions depend heavily upon the significance of the "synaptic activation noise" recorded during muscle stretch and on the assumption that changes in firing rates of artificially depolarized neurons reflect only postsynaptic actions, and, therefore, these results are in need of further confirmation. Nevertheless it may be postulated that the strychnine-resistant, picrotoxin-sensitive postsynaptic inhibition occurs in parallel to presynaptic inhibition. This possibility is suggested by the histological findings (see Figs. 7 and 9) that the presynaptic parts of axo-axonic synapses often also have specific synaptic contacts with those neurons on which the afferent terminals end. At the neuromuscular junction of the crayfish pre- and postsynaptic inhibition are always exerted by the same inhibitory fibre (Dudel and Kuffler, 1961).

2.5.3 Cholinergic Drugs

The PAD evoked both in the cat and in the toad spinal cord by peripheral nerve or dorsal root stimulation was not or only very little affected by a wide variety of drugs active at cholinergic synapses (Kiraly and Phillis, 1961; Eccles, Schmidt and Willis, 1963 f). Thus, dihydro-β-erythroidine, gallamine triethiodide (Flaxedil), atropine, nicotine, tetraethyl-pyrophosphate, and eserine had no action on PAD or presynaptic inhibition when injected intravenously or applied topically in concentrations highly active on known cholinergic synapses. In high concentrations acetylcholine, carbachol, succinylcholine, d-tubocurarine and nicotine decreased the amplitude of the amphibian cord DRP induced by dorsal root stimulation (Barron and Matthews, 1938; Eccles, 1947; Kiraly and Phillis, 1961; Grinnell, 1966; Phillis and Tebécis, 1967; Koketsu et al., 1969), but this may be considered a non-specific effect of these drugs, particularly as their actions were frequently irreversible. Therefore, it seems unlikely that the pathways of PAD from sensory fibres to sensory terminals contain a cholinergic synapse.

An interesting exception to this general rule is the DRP that in the amphibian spinal cord results from antidromic stimulation of ventral roots (Barron and Matthews, 1938; Eccles and Malcolm, 1946; Lloyd and McIntyre, 1949). The latter authors demonstrated that this potential, usually designated VR-DRP, although being of smaller magnitude and having a longer latency, corresponded in general outline to the DRP (more specifically the DRP V) set up by stimulation of an adjacent dorsal root. The VR-DRP is reduced in amplitude by acetylcholine (Eccles, 1947), and Kiraly and Phillis (1961) have shown that it can also be depressed by dihydro-β-erythroidine, and to a lesser extent by atropine, curare and hexamethonium. Anticholinesterase substances in low concentrations potentiated it (see also Koketsu, 1956a) and in higher concentrations depressed it. Similar results were reported by Grinnell (1966). These findings indicate that the reflex pathway from motor-axon collaterals to afferent fibre terminals contains a cholinergic synapse. Presumably this synapse corresponds to the synapse made by the motor-axon collaterals at the Renshaw cells in the cat spinal cord. It has been shown that acetylcholine and carbachol when directly applied to the amphibian spinal cord produced a slow depolarization of afferent fibre terminals and it has been pointed out that the pharmacological properties of this depolarization were very similar to those of the VR-DRPs (Koketsu et al., 1969). These effects are most probably not due to a direct action of these drugs on the fibre ter-

minals, but reflect the activation of the cholinergic receptor sites of the interneurones on the VR-DRP pathway.

2.5.4 Amino Acids

A wide variety of amino acids is known to increase or decrease the excitability of postsynaptic elements and several of the amino acids have been
proposed as possible transmitter substances (for reviews see CURTIS, 1963,
1968, 1969). Among the most potent members of the depressant group are
the neutral amino acids, γ-amino-butyric acid (GABA) and 3-amino-1-propane-
sulphonic acid. In crustaceae GABA has an action which is identical with the
inhibitory transmitter substance at several types of inhibitory synapses (for
literature see FURSHPAN, 1959; GRUNDFEST et al., 1959; FLOREY, 1961, 1964;
CURTIS, 1963; POTTER, 1968). This similarity also obtains for presynaptic
inhibitory action (DUDEL and KUFFLER, 1961). Picrotoxin acts as a blocking
agent at these synapses, depressing both the inhibitory synaptic action and
the action of GABA (ELLIOTT and FLOREY, 1956; ROBBINS and VAN DER KLOOT,
1958; VAN DER KLOOT, 1960; FLOREY, 1961, 1964).

Since presynaptic inhibition in the cat is reversibly antagonized by picrotoxin, the effect of topically administered GABA and 3-amino-1-propane-
sulphonic acid on PAD has been tested by ECCLES, SCHMIDT and WILLIS
(1963 f). It was found that both substances reduced and shortened the DRPs
and increased the dorsal root reflexes. Both amino acids also depressed the
DRP in the isolated toad spinal cord (PHILLIS, 1960; SCHMIDT, 1963; TEBÉCIS
and PHILLIS, 1969a). Simultaneously they increased the excitability of the
primary afferent fibres. However, the excitability increases were much smaller
than would have been expected from the large DRP reduction. Furthermore,
CURTIS and RYALL (1966) tested the action of electrophoretically administered
GABA upon the excitability of the preterminal regions of spinal afferent fibres
and found that under these much more defined experimental conditions GABA
did not increase but actually decreased the electrical excitability of the terminals, and this has recently been confirmed on cuneate fibre terminals
(GALINDO, 1969). These results underline the original assumption of ECCLES,
SCHMIDT and WILLIS (1963 f) and SCHMIDT (1963) that topically administered
GABA (and γ-amino-propane-sulphonic acid) reduced the DRP not, or not
only, by depolarizing the presynaptic fibre terminals but also by actions at
other sites of the PAD pathway. Certainly, further clarification of the situation
is needed, particularly since GALINDO (1969) reported that the pre- and postsynaptic depressant actions of GABA could be fully blocked by picrotoxin.

Acidic amino acids such as glutamic acid excite postsynaptic elements.
CURTIS et al. (1961) found that glutamic acid increased the ventral root reflex
discharge of isolated toad spinal cords and that it depolarized motoneurones.
PHILLIS (1960) reported that in the same preparation glutamic acid reduced

the DRP evoked by dorsal root stimulation and this finding has been confirmed (Schmidt, 1963; Phillis and Tebécis, 1967; Tebécis and Phillis, 1967, 1969a, b). A comparison of the DRP depression and the excitability increase of primary afferent fibres showed that, unlike GABA and 3-amino-1-propane-sulphonic acid, glutamic acid probably directly depolarized the primary afferent fibres. Curtis and Ryall (1966) in their microelectrophoretic study also found excitability increases of presynaptic fibres upon glutamic acid (or DL-homocysteic acid) ejection. Nevertheless the significance of the results obtained with acidic amino acids remains just as obscure as that of the changes measured upon GABA administration.

2.5.5 Catecholamines

Except for the specific blocking action of dihydro-β-erythroidine on the DRP evoked in the amphibian spinal cord by ventral root stimulation, all pharmacological effects reported so far have been found in all types of PAD

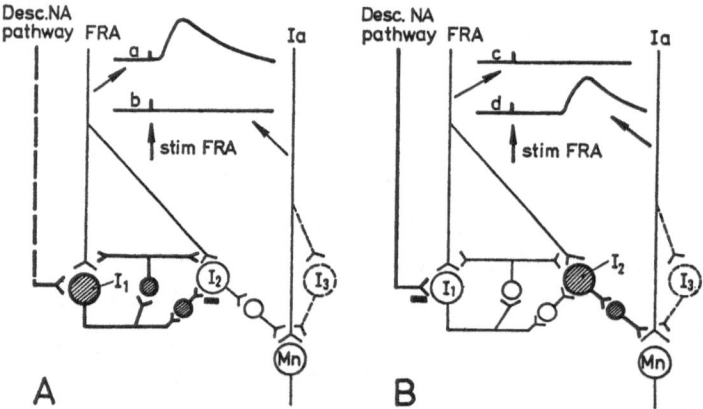

Fig. 21 A and B. Influence of descending noradrenergic pathway on PAD pathways. The diagrams were constructed to illustrate the findings and conclusions of Lundberg and collaborators quoted in the text. A gives the situation in the acute spinal cat, B that in a cat with an intact descending noradrenergic pathway. For further discussion of the diagrams see text

independent of their origin. Meanwhile, there are also reports of drug actions affecting only one or the other system of PAD. Lundberg and his co-workers found that the PAD induced in ipsi- and contralateral flexor reflex afferents (FRA) by volleys in other FRA were markedly reduced after intravenous injection of DOPA (L-3,4-dihydroxy-phenylalanine), whereas the DRPs induced in Group I muscle afferents were not affected. (The transmission from the FRA to motoneurones and ascending pathways was also depressed). Under these conditions volleys in FRA evoked a long latency PAD in ipsi- and contralateral Group Ia fibres from flexors and extensors (together with a powerful long latency flexor reflex). A detailed physiological and pharmacological analysis of these phenomena revealed that, most probably, descending inhibitory noradrenergic tracts end on interneurones of the FRA interneuronal pathways, and that DOPA mimicks activity in these descending tracks by inducing synthesis and overflow of noradrenaline from their synaptic terminals (ANDÉN, LUNDBERG, ROSENGREN and VYCKLICKY, 1963; ANDÉN, JUKES and LUNDBERG, 1964, 1966; ANDÉN, JUKES, LUNDBERG and VYCKLICKÝ, 1964, 1966a, b; JANKOWSKA et al., 1966; LUNDBERG, 1966; ENGBERG et al., 1968).

The diagrams in Fig. 21 show schematically the proposed pathways of FRA fibres to the presynaptic terminals of FRA and muscle Group I a afferent fibres. The situation in the acute spinal cat is given in A. A volley in the FRA fibres evokes activity in the interneuronal chains starting at the interneuron I_1 and this activity results in PAD of FRA fibres (DRP specimen a in A), and in inhibition of the pathway starting at interneurone I_2 and leading to the I a preterminals. No PAD is induced in these terminals (specimen b in A). Activation of the descending inhibitory pathway (NA) or injection of DOPA results in the situation shown in B; the interneuronal chain I_1 is inhibited. Thus no PAD can be recorded from FRA fibres (specimen c) but the I a fibres are depolarized via I_2 (specimen d). The pathway I_3 leading from Group I a afferents to Group I a afferents remains unaffected.

The existence of descending noradrenergic pathways and the absence of segmental noradrenergic neurones have been suggested by the use of histochemical and biochemical methods (CARLSSON et al., 1964; DAHLSTRÖM and FUXE, 1965). The proposal that DOPA acts not directly on PAD interneurones but by liberating transmitter (most probably noradrenaline) from such descending pathways is largely based on pharmacological evidence. First of all DOPA has little or no effect after inhibition of DOPA decarboxylase (by meta-hydroxy-benzylhydrazine), and its effect is potentiated by monoaminooxidase inhibition (by nialamide). These findings suggest that DOPA acts only after decarboxylation (to dopamine which is converted to noradrenaline) and not directly. Furthermore, reserpine pretreatment prevents the effects of DOPA. Finally the effects of DOPA can be antagonized by the blockers of adrenergic α-receptors, phenoxybenzamine and chlorpromazine, but not by the β-blocker pronethalol (Nethalide).

Injection of 5-HTP (5-hydroxytryptophane, precursor of 5-HT, 5-hydroxytryptamine) has actions similar to those seen after DOPA injection (ANDÉN, JUKES and LUNDBERG, 1964). As summarized by LUNDBERG (1966) there is pharmacological evidence that, as with DOPA, effects of 5-HTP are due to liberation of 5-HT from terminals of a descending pathway, and that these terminals have receptor sites different from those activated by DOPA.

2.5.6 Varia

The effect of tetrodotoxin (TTX) on DRPs generated by dorsal root stimulation or by direct stimulation of the dorsal horn was studied by RUDOMIN and MUNOZ-MARTINEZ (1969). It was shown that the DRP produced by dorsal root stimulation completely disappeared upon intra-arterial TTX administration (40–80 µg/kg), whereas the DRP generated by dorsal horn stimulation was depressed but not abolished. Several explanations of this phenomenon were offered by the authors. To them the most likely explanation appeared to be that, by analogy with the effects of depolarizing currents at TTX poisoned synapses and neuromuscular junctions (BLOEDEL et al., 1966, 1967; KATZ and MILEDI, 1967a–c), the stimulating current released transmitter from the presynaptic terminals of axo-axonic synapses and that this transmitter evoked the observed TTX resistant DRPs.

CARLSSON (1964) tested the effects of Na^+-reduction on the DRP. Frog spinal cords were perfused with Ringer's solution (normal colloid osmotic pressure achieved by addition of dextran) and the Na^+-concentration of the perfusion fluid was varied. The DRP was shown to vary linearly with the log_{10} of the Na^+-concentration (Na^+ replaced by sucrose). These results were interpreted as evidence for the rôle of an internuncial system in the generation of the DRP.

Different types of prostaglandins depolarize amphibian dorsal root fibres and reduce the size of DRPs (PHILLIS and TEBÉCIS, 1968). This effect has an extremely slow rate of recovery (0.5–1 h). The site of action is unknown. Increases in the H^+-concentration caused a depolarization of dorsal root terminals and a simultaneous decrease in amplitude and increase in duration of the DRP (TEBĒCIS and PHILLIS, 1968).

3. Functional Organization of Presynaptic Inhibition in the Mammalian Central Nervous System

Introductory Remarks. In the studies concerned with the functional organisation of presynaptic inhibition the various manifestations of preterminal depolarization have usually been used as indicators for presynaptic inhibitory processes. The inhibition itself was rarely measured. It was generally assumed that the amount and time course of the preterminal depolarization was directly related to the amount and time course of the presynaptic inhibition exerted onto the depolarized fibres. (For the justification and the limitations of this assumption see section 2). Thus, strictly speaking, most of the research reviewed here was concerned with the central organization of the reflex pathways leading either to a primary afferent preterminal or to another axonic preterminal. Therefore, the material has been arranged in relation to the various fibre types which undergo depolarization.

Two further general comments should be made before starting a detailed consideration of the organization of PAD. First of all, it should be mentioned that practically all results reported here have been obtained from the central nervous system of the cat. It has to remain open, therefore, to what extent the organization of the PAD reflex pathways in other mammals resembles that described here. Secondly it should be made clear that on the one hand it is usually possible to discover the relative amounts of presynaptic depolarization exerted from various inputs onto a given fibre population, whereas on the other hand, when a given input acts on different fibre populations it is difficult to judge the relative amounts of the respective depolarizations. Comparisons of the latter type can only be made when the depolarization is measured directly by intrafibre recording, and this has only been done systematically when testing the relative potency of spinal afferents in depolarizing other spinal afferents (ECCLES, MAGNI and WILLIS, 1962; ECCLES, SCHMIDT and WILLIS, 1963 a, e).

3.1 Presynaptic Inhibition and PAD Induced in Muscle Afferents of the Spinal Cord

3.1.1 PAD Induced in I a Afferents

PAD by Segmental Inputs. FRANK and FUORTES (1957) in their first investigation of presynaptic EPSP depression used conditioning volleys from ipsilateral hamstring nerves to depress the monosynaptic EPSPs of GS moto-

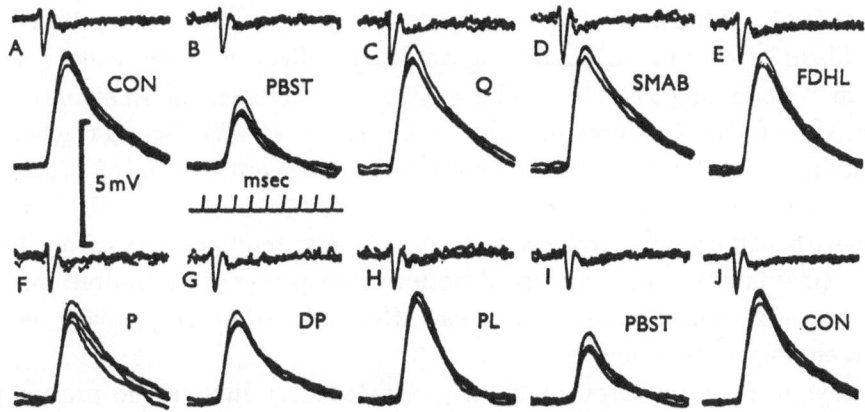

Fig. 22A–J. Presynaptic inhibition of monosynaptic EPSPs induced by activity in different muscle afferents. The EPSPs were generated in a gastrocnemius motoneurone by maximum Group I GS volleys at a repetition rate of 5/s and each record is formed by the superposition of 3 or 4 traces. A and J are control records. In B–I various muscle nerves, as indicated by the symbols, were stimulated at 300/s and just supramaximally for Group I. The EPSP were recorded about 1 s after the onset of the tetani (ECCLES, ECCLES and MAGNI, 1961)

neurones (cf. also Fig. 1). ECCLES, ECCLES and MAGNI (1961) showed that the flexor component of these nerves (PBST) was particularly potent in depressing monosynaptic EPSPs both of flexor and extensor motoneurones. Furthermore, they found that other flexor nerves of the leg (P, DP) gave comparable amounts of EPSP depression, whereas several kinds of extensor nerves were much less effective (Fig. 22). The EPSP depression was produced by volleys both in Group Ia (from primary muscle spindle endings) and Group Ib (from Golgi tendon organs) afferent fibres, and little if at all by Group II (from secondary muscle spindle endings) and Group III afferent impulses. A similar pattern was observed when testing the presynaptic inhibition of spinal monosynaptic reflex pathways (Eccles, Schmidt and Willis, 1962; Decandia, Gasteiger and Mann, 1968) and when measuring the presynaptic EPSP depression in the cervical spinal cord (Schmidt and Willis, 1963a). The presynaptic inhibition was exerted both onto those Ia terminals making monosynaptic excitatory contacts on motoneurones and onto those conveying excitation to neurones of ascending tracts (Eccles, Schmidt and Willis, 1963d; Jankowska et al., 1964, 1965).

It has to be concluded that Ia fibres from primary endings of muscle spindles of flexor and extensor muscles are presynaptically inhibited, i.e. depolarized, by volleys mainly in Ia and Ib fibres of (ipsilateral) flexor muscles. This conclusion was fully confirmed in the subsequent investigations dealing with the PAD exerted onto Ia fibres (Fig. 23) by various ipsilateral segmental inputs (Eccles, Kozak and Magni, 1961; Eccles, Magni and Willis, 1962; Eccles, Schmidt and Willis, 1963a, b; Eccles and Willis, 1962; Schmidt and Willis, 1963b; Voorhoeve and Verhey, 1963; Cook et al., 1965; Verhey et al., 1966; Gillies et al., 1969; Barnes and Pompeiano, 1970a–d). Volleys in Ib fibres often proved to be slightly more effective than volleys in Ia fibres in depolarizing Ia fibres. The greater effectiveness of flexor Ib fibres compared to flexor Ia fibres in depolarizing Ia fibres was also seen when the presynaptic inhibitory effects upon stretch and contraction of ipsilateral muscles were studied (Devanandan et al., 1964, 1965a, b, 1966). Under conditions with particularly strong activation of Golgi tendon organs from flexor muscles (maximum stretch and contraction), the presynaptic inhibition of Ia fibres exceeded considerably that seen after activation of primary muscle spindle endings only (slight stretch).

Volleys in extensor Group I fibres generated very little or no presynaptic inhibition in Ia fibres of flexor muscles. However, contraction of extensor muscles did induce PAD in agonist extensor muscles (Devanandan, Eccles and Stenhouse, 1966). Experiments with electrical stimulation of extensor nerves (Decandia et al., 1966) or activation of primary spindle endings by vibratory stimuli (Gillies et al., 1969; Barnes and Pompeiano, 1970b–d) confirmed that stimulation of Group I agonist fibres depolarized Ia extensor terminals, and evidence was presented that this depolarization was paralleled by a presynaptic inhibition of the agonist monosynaptic reflex (Decandia et al., 1967; Gillies et al., 1969; Barnes and Pompeiano, 1970b–d; cf. also Eccles, Schmidt and Willis, 1962). Usually, repetitive activation had to be used to obtain appreciable effects, single afferent volleys being ineffective

(WALL, 1958). The results indicate that the PAD produced by extensor Group I fibres has a more circumscribed feedback character than that evoked by activity in flexor Group I fibres, although there seem to be some exceptions to this rule (cf. ECCLES, SCHMIDT and WILLIS, 1963 d).

Several types of muscle and cutaneous afferent volleys have proved to be ineffective in the spinal animal in evoking PAD in Group Ia afferent fibres. First of all, all types of contralateral afferent volleys from muscle and skin nerves had no depolarizing effect on Ia afferents and did not depress the

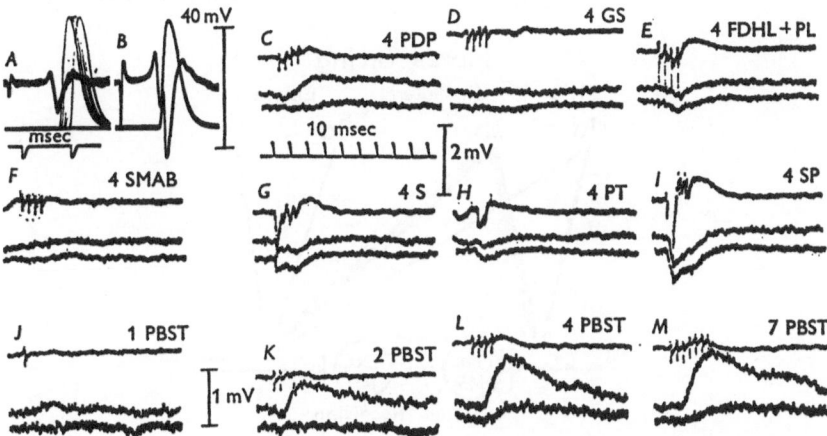

Fig. 23 A–M. PAD induced in a Ia fibre by muscle and cutaneous afferent inputs. Intracellular recording from a quadriceps Ia fibre as shown by A and B. In all other records (C–M) the upper trace records the afferent volleys, the second trace is the intracellular potential, while the lowest trace gives the potential produced by an identical series of nerve volleys, but recorded after withdrawal of the microelectrode to a just extracellular position. Muscle afferent volleys were maximum for Group I, while cutaneous volleys were generated by stimuli three to four times threshold (ECCLES, MAGNI and WILLIS, 1962)

size of Ia EPSPs (DEVANANDAN et al., 1965). Secondly, volleys in ipsilateral flexor reflex afferents, FRA, i.e. myelinated cutaneous afferents (Group II fibres), and high threshold muscle afferents (Group II and III fibres), did not induce PAD in Ia fibres (ECCLES, KOZAK and MAGNI, 1961; ECCLES, MAGNI and WILLIS, 1962; ECCLES, KOSTYUK and SCHMIDT, 1962a; LUND, LUNDBERG and VYCKLICKÝ, 1965). They did, however, inhibit the PAD evoked in Ia afferents by volleys in Ia afferents (LUND et al., 1965). This is one of several examples indicating that afferent activity not only produces PAD, but is also able to inhibit it. This inhibition presumably takes place on the interneurones of the PAD pathway, possibly presynaptically through depolarization of interneuronal terminals.

After injection of DOPA into spinal animals, previously ineffective volleys in FRA evoked a long latency PAD in ipsi- and contralateral Group Ia fibres from flexors and extensors. The possible descending pathways responsible for these effects were carefully investigated by LUNDBERG and his collaborators. They are described in detail in chapter 2.5.5 and diagrammatically summarized in Fig. 21. The results point to the considerable influence exerted by supraspinal pathways onto the excitability and hence the excitatory synaptic efficacy of primary afferent fibres.

The summarizing diagram of Fig. 24 shows schematically the different afferent inputs producing PAD of extensor and flexor Ia terminals, respectively. The thickness of the arrows gives an estimate of the potencies of these various pathways in producing PAD. These estimates are based on an appreciation of the results reviewed in chapter 3.1.1 and indicate merely that,

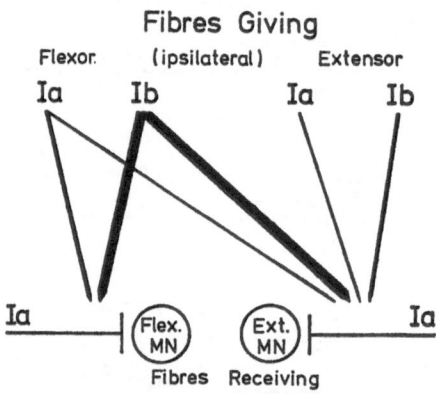

Fig. 24. Ipsilateral inputs depolarizing Ia fibres of flexor and extensor muscles. The approximate relative amount of depolarization contributed by each input has been estimated from the results quoted in the text and is indicated by the width of the arrow. MN, motoneurone

according to the evidence presently available, not all types of Group I muscle afferents are equally potent in producing PAD of either flexor or extensor Ia fibres.

PAD of Ia Afferents by Suprasegmental Inputs. Stimulation of the ipsi- or contralateral sensorimotor cortex or of the pyramidal tract did not evoke PAD in Ia afferents (Andersen, Eccles and Sears, 1962, 1964a; Carpenter, Lundberg and Norrsell, 1962, 1963). However, Carpenter, Engberg and Lundberg (1962, 1966) found in a systematic investigation of the PAD evoked by repetitive stimulation of the brain stem and of the cerebellum that upon stimulation of a dorsal midline region of the medulla a depolarization was induced in lumbar Ia terminals of flexor and extensor muscles. This PAD disappeared after the transection of the ventral quadrants of the spinal cord. It was concluded that this PAD was due to the activation of the ipsilateral medial longitudinal fasciculus.

Similar effects were obtained by Cook et al. (1969b) when stimulating the ipsilateral VIIIth cranial nerve. They attributed these effects to the vestibular nuclei and their descending afferent projection pathway, because the same results were obtained when stimulating the vestibular complex, especially the medial and descending vestibular nuclei and the medial longitudinal fasciculus at mesencephalic levels. Further, complete lesions of the vestibular nuclei abolished the DRPs evoked by stimulation of the VIIIth nerve (COOK et al., 1969a). Stimulation of the contralateral caudal part of the closely related fastigial nucleus also elicited PAD in Ia fibres of flexor and extensor muscles (CANGIANO et al., 1969a). The PAD induced by the various types of brain stem stimulation was practically always associated with a depression of the monosynaptic reflex to flexor muscles, whereas the monosynaptic reflex to extensors was often facilitated due to the concurrent excitation exerted on extensor motoneurones by stimulation of the brain stem, the VIIIth nerve, and the vestibular and fastigial nuclei.

The significance of the results described in the last two paragraphs for the regulation of movement and posture is difficult to evaluate (cf. LUNDBERG, 1966, 1967). Generally it appears that (unlike Ib and cutaneous afferents, see below) Ia fibres cannot be depolarized by activation of cortical areas, whereas some subcortical descending pathways seem to be quite effective. Most probably further investigations will reveal additional and even more complex pathways involved in the control of the excitatory input from Ia afferents. For instance in the unrestrained, unanaesthetized cat, a PAD of Ia terminals has been observed during the phasic events of the desynchronized phase of sleep (MORRISON and POMPEIANO, 1965, 1966; BALDISSERA et al., 1966; BALDISSERA and BROGGI, 1967). Possibly this PAD is responsible for the transient depression of the homonymous monosynaptic reflexes that occurs during these periods of desynchronized sleep.

3.1.2 PAD Induced in Ib Afferents

PAD by Segmental Inputs. Using a wide variety of PAD detecting techniques ECCLES, SCHMIDT and WILLIS (1963a) established that the main ipsilateral segmental sources from which flexor and extensor Ib endings were depolarized were the Ib fibres both of flexor and extensor muscles. The different pattern of Ia versus Ib fibre depolarization becomes evident when comparing Fig. 25 with Fig. 23 (for comparison with the corresponding cutaneous PAD patterns see Figs. 3 and 4). It is particularly striking that the extensor Group I volleys were just as effective in depolarizing Ib fibres as the flexor Group I volleys. By various stimulating and recording procedures it was established that volleys in Ia fibres of all types of muscle nerves did not produce any PAD in Ib fibres, so that the effects of Group I volleys seen in Fig. 25 have to be attributed to the activity in Ib afferents. There were, however, definite contributions to the PAD of Ib fibres when high threshold muscle afferents (Group II and III fibres) were included in the volley or when cutaneous nerves were stimulated. The predominance of Ib fibres in depolari-

5*

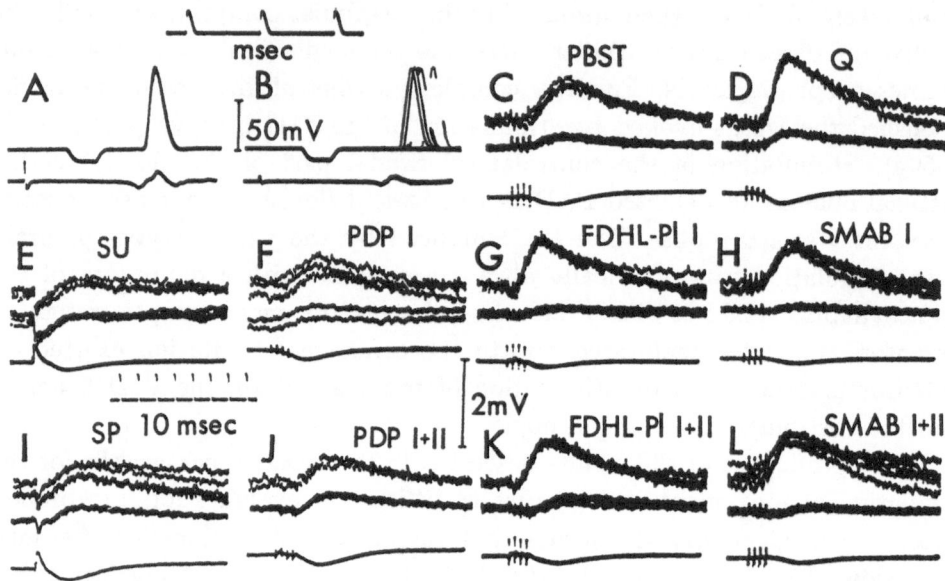

Fig. 25 A–L. PAD induced in a I b fibre by muscle and cutaneous afferent inputs. Intracellular recording from a GS I b fibre; the identification of the fibre type is shown in A and B. In C–L the upper traces show the intracellular recordings, the middle traces the extracellular field potentials produced by the same volleys, and the lower traces the cord surface potentials. In C, D and F–H the muscle nerves were stimulated four times (300/s) at maximal Group I strength. In J–L the stimulus strength was maximal for both Groups I and II. The cutaneous nerves in E and I were stimulated once at a strength of 4 times threshold
Eccles, Schmidt and Willis, 1963a)

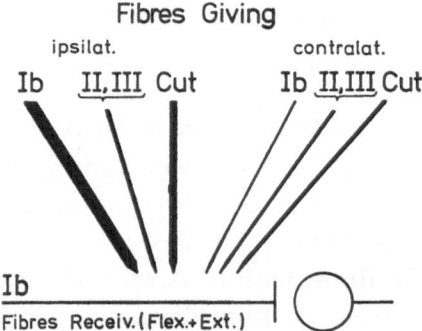

Fig. 26. Ipsi- and contralateral inputs depolarizing I b fibres. The approximate relative amount of depolarization contributed by each input has been estimated from the results quoted in the text and is indicated by the width of the arrow. I b, II and III are the respective muscle afferents, Cut represents the myelinated cutaneous afferents

zing I b afferents was confirmed in experiments evoking activity in I a and I b fibres by muscle stretch (Devanandan et al., 1965a).

The left-hand side of the diagram in Fig. 26 summarizes the results obtained by Eccles, Schmidt and Willis (1963a). Here, as with the PAD of I a fibres, it has not been possible to discover any topographical relationship. In view of this failure, the flexor and extensor I b fibres receiving the depolarization were pooled together at the bottom of the figure, and the afferent inputs were assembled to give the basic types Groups I b, II, III, and cutaneous. As in Fig. 24 the thickness of the arrows gives an approximate measure of the potencies of these various pathways in producing PAD.

Contralateral volleys in certain types of afferent fibres also depolarize I b terminals. DEVANANDAN et al. (1965) showed that the crossed PAD of I b fibres originated from high threshold (II and III) muscle afferents and from cutaneous fibres. Their results also indicated that I b fibres have a weak depolarizing action onto the contralateral I b nerve terminals. These results are also included in Fig. 26. Again, the relative effectiveness of these inputs is given by the thickness of the arrows. It has to be pointed out, however, that these estimates are based on relatively indirect evidence, and can only be considered as first approximations.

PAD of I b Afferents by Suprasegmental Inputs. In parallel and independent work, it was recognized by ANDERSEN, ECCLES and SEARS (1962, 1964a) and by CARPENTER, LUNDBERG and NORRSELL (1962, 1963) that stimulation of the sensorimotor cortex produced PAD of I b (and Group II and cutaneous) afferent fibres. The effective cortical areas were the somatosensory areas I and II. The effect was predominantly contralateral from S I, whereas the S II effect was bilateral. The S I arm and leg areas acted specifically on the arm and leg afferent fibres, respectively. Single cortical stimuli were relatively ineffective, whereas repetitive stimulation at high frequency caused a large recruitment of the responses. This summation took place in the spinal cord, not at the cortical level. The effects were shown to be mediated by pyramidal and by extraspinal pathways (HONGO and JANKOWSKA, 1967; HONGO and OKADA, 1967). Spatial facilitation between the paths from primary afferents and cortex was observed: cortical stimulation at a strength subliminal for effects on primary afferents markedly facilitated submaximal dorsal root potentials evoked from Group I afferents. In addition, interneurones of the D-type (cf. chapter 2.4.2) were found in the dorsal horn and in the intermediate nucleus of CAJAL that were fired by cortical stimulation and otherwise had properties making them possible candidates for the mediation of the cortically evoked PAD. All these findings indicate that an interneuronal system mediates the transmission from the synaptic terminals of the cortico-spinal fibres to the presynaptic inhibitory endings on the synapses of the primary afferent fibres.

Repetitive electrical stimulation of wide areas of the brain stem of decerebrate cats also evoked considerable PAD of I b fibres (CARPENTER, ENGBERG and LUNDBERG, 1962, 1966). Two main regions were recognized. One is a very distinct area in the midline region about 1 mm ventral to the 4th ventricle in the medulla. Stimulation of this area also produced PAD in I a and cutaneous fibres. As already outlined in chapter 3.1.1 (Suprasegmental Inputs), it was concluded that this PAD was due to the activation of the ipsilateral medial longitudinal fasciculus. The second area is a more widespread ventromedial region. Stimulation of this area evoked PAD in Ib and cutaneous fibres only. These effects may be due in part to the activation of pathways descending from higher centers, but it is interesting to note that the lowest thresholds for evoking these effects were found in a region corresponding to Magoun's inhibitory center.

From other ventro-medial regions of the caudal brain stem, it was possible to evoke positive DRPs, indicating primary afferent hyperpolarization, PAH, i.e. presynaptic disinhibition, and to depress negative DRPs evoked from primary afferents (Lundberg and Vycklický, 1963, 1966). Vice versa, tonic presynaptic depolarization produced by brain stem stimulation was temporarily interrupted by volleys in FRA fibres. Evidence was presented by Lundberg and Vycklický (1966) that these manifestations of PAH reflected inhibition of tonic presynaptic depolarizations, i.e. the temporal removal of tonic presynaptic inhibition. (At the segmental level, examples of PAH have been reported by Mendell and Wall, 1964; Lund et al., 1965; Mendell, 1970; and Dawson et al., 1970).

The complex PAD effects which can be elicited from the VIIIth cranial nerve and the vestibular and fastigial nuclei have already been reported (chapter 3.1.1, Suprasegmental Inputs, cf. Cook et al., 1968, 1969a, b; Cangiano et al., 1969a). In decerebrate cats, stimulation of the lateral parts of the intermediate region on both sides of the anterior cerebellar cortex also induced large PAD of Ib (and cutaneous) afferents (Carpenter et al., 1966). Cangiano et al., (1969b) confirmed these results and added the observation that, in some experiments, stimulation of the medial part of the vermal cortex of the cerebellar anterior lobe elicited positive DRPs, i.e. PAH. By stimulation and ablation experiments they were able to show that this PAH reflected a tonic inhibitory control of the cerebellar cortex on the vestibular nuclei, which in turn generated a tonic PAD of Group I as well as cutaneous afferents. (No effort was made to distinguish between effects on Ia and Ib afferent fibres).

3.1.3 PAD Induced in Group II and III Afferents from Muscle

Very little is known of the conditions which may induce PAD in myelinated high threshold muscle afferents (Group II and III afferents). It is generally assumed that these fibres not only contribute to the FRA system (as defined by Eccles and Lundberg, 1959), thus evoking PAD in Ib and cutaneous afferents (cf. Eccles, Kostyuk and Schmidt, 1962a, b; Eccles, Schmidt and Willis, 1963a, e), but also receive PAD from flexor reflex afferents. The only direct evidence to support this assumption has been reported by Eccles,

SCHMIDT and WILLIS (1963a), who recorded intracellularly from a small number of Group II fibres the PAD induced by cutaneous and muscle nerve stimulation. The results suggested that Group II muscle afferents are more depolarized by cutaneous volleys than by Group I muscle volleys. Descending systems also seem to be able to evoke PAD in Group II afferents: CARPENTER et al. (1963) observed that, of 12 identified muscle Group II fibres, 6 displayed an increased excitability after stimulation of the sensorimotor cortex.

3.2 Presynaptic Inhibition and PAD Induced in Cutaneous Afferents of the Spinal Cord

3.2.1 PAD Induced by Electrical Stimulation of Spinal Nerves and Supraspinal Structures

PAD Induced by Spinal Nerves. In the early investigations on PAD, it was soon recognized that volleys in myelinated cutaneous afferents were particularly powerful in evoking a large PAD of spinal afferents. Indirect methods, such as the recording of dorsal root reflexes (TOENNIES, 1938, 1939) and the testing of the excitability of the spinal axon terminals (WALL, 1958), indicated that

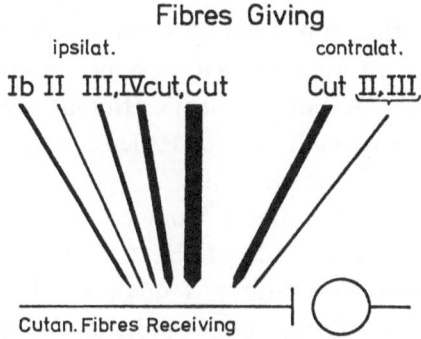

Fig. 27. Ipsi- and contralateral inputs depolarizing cutaneous afferent fibres. The approximate relative amount of depolarization contributed by each input has been estimated from the results quoted in the text and is indicated by the width of the arrow. Ib, II and III are the respective muscle afferents, Cut represents the myelinated cutaneous afferents and IVcut the unmyelinated cutaneous afferents

this PAD was mainly exerted onto the cutaneous fibres themselves, both those that have been primarily activated and those initially passive. These original observations were fully confirmed by the subsequent intrafibre studies of cutaneous PAD (KOKETSU, 1956b; ECCLES and KRNJEVIĆ, 1959a, Table 2; ECCLES, SCHMIDT and WILLIS, 1963e, Table 1). A typical example for such measurements is illustrated in Fig. 3.

Apart from the predominant influence of cutaneous afferents onto the cutaneous PAD system, several other *ipsilateral* segmental inputs are able to depolarize cutaneous fibres. First of all, activation of muscle Ib fibres, both of flexor and extensor muscles, usually gave appreciable PAD in cutaneous fibres (cf. Fig. 3), whereas stimulation of Ia fibres almost always remained completely ineffective (Eccles, Schmidt and Willis, 1963 e, Fig. 5). Secondly, Group II and III fibres from muscle, which in many respects work in synergy with cutaneous fibres, also exert a considerable depolarizing influence on cutaneous afferent terminals (Eccles, Kostyuk, Schmidt, 1962a, b; Eccles, Schmidt and Willis, 1963 e). Thirdly, volleys in cutaneous unmyelinated fibres (C-fibres) lead to depolarizations of cutaneous myelinated fibres which are often just as powerful as those induced by cutaneous A-fibres (Jänig and Zimmermann, 1971). Finally, Selzer and Spencer (1969b) showed that stimuli to the visceral afferents in the caudal sympathetic chain induced PAD in A delta cutaneous afferents of the lumbar cord (and vice versa, see chapter 3.5). The left-hand side of Fig. 27 summarizes diagramatically the ipsilateral segmental somatic inputs giving PAD of cutaneous myelinated afferent fibres. Except for arrow IVcut. the thicknesses of the arrows are proportional to the average PAD measured by Eccles and Krnjević (1959a) and by Eccles, Schmidt and Willis (1963 e) in a population of about 100 fibres. The thickness of arrow IVcut. was taken from intracellular measurements of Jänig and Zimmermann (1971) who compared the PAD produced by sural A-volleys with that produced by sural C-volleys during A-fibre block. Thus the diagram gives an approximate measure of the relative potencies of the various inputs giving PAD of cutaneous afferent fibres. It should be added in parentheses that so far no evidence for PAD of cutaneous or muscle afferent C-fibres has been brought forward.

The observations on PAD effects from contralateral afferent fibres are summarized on the right-hand side of Fig. 27 from the results reported by Eccles, Holmqvist and Voorhoeve (1964a, b). In general, the features displayed by volleys in contralateral cutaneous and muscle nerves onto cutaneous afferents resembled those evoked from the ipsilateral side. However, the contralateral effects were always much weaker and, as always with contralateral actions, their latency was longer. Furthermore, when contralateral muscle nerves were stimulated, volleys in Ia and Ib fibres remained ineffective while Group II and III volleys evoked a presynaptic depolarization.

PAD Induced by Supraspinal Structures. It has been a general finding that those supraspinal structures exerting PAD on muscle Ib fibres almost always also depolarized spinal cutaneous afferents. The reader is referred, therefore, to the last 4 paragraphs of chapter 3.1.2 where the relevant papers have been reviewed. PAD effects from supraspinal structures restricted to cutaneous afferents have not been described so far. In a short note Calma (1966) reported that stimulation of the ventral thalamo-diencephalic region caused bilateral PAD of cutaneous afferents, but he did not specifically state that a simultaneous PAD of muscle Ia or Ib afferents had been excluded.

3.2.2 PAD Induced in Spinal Cutaneous Afferents by Adequate Stimulation of Skin Receptors

Mechanical Stimuli. When electrical stimulation of cutaneous nerves was used, the question remained undecided as to which sensory modalities produced and received PAD. To answer these questions two modifications of the experimental technique were necessary: the PAD of single afferent fibres of known modality had to be measured within the spinal cord (Fig. 5), and electrical stimulation of cutaneous nerves had to be replaced by natural stimulation of skin receptors.

In detailed studies of the PAD produced in different types of mechanoreceptor afferents by carefully controlled mechanical stimuli to the cat's foot

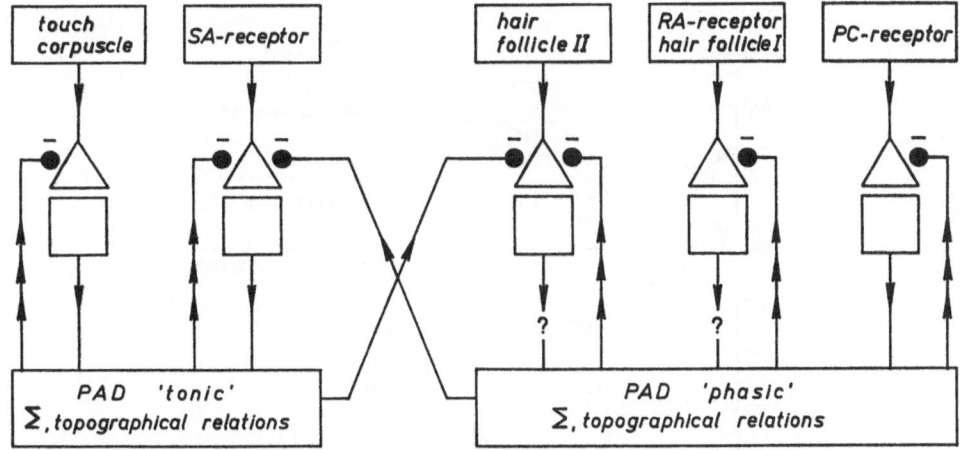

Fig. 28. The operational relationships of the cutaneous PAD pathways. Schematic diagram illustrating the presynaptic inhibitory connections within the cutaneous mechanoreceptor afferent population. The boxes in the upper row stand for the various receptor types, the slowly adapting receptors being assembled on the left-hand side. The middle row complex indicates the first synapses in the spinal cord onto which the receptors project, the closed circles showing the presynaptic control. The PAD generating systems are shown in the lower row. The number of arrows in the feedback lines denote the feedback gain. SA and RA designate slowly and rapidly adapting receptors respectively. PC are Pacinian corpuscles (JÄNIG, SCHMIDT and ZIMMERMANN, 1968b)

(cf. JÄNIG et al., 1968a) it was found that two separate systems generate PAD in cutaneous afferents, both being of negative feedback character (JÄNIG et al., 1968b). As schematically shown in Fig. 28, one system is activated by impulses from rapidly adapting low threshold receptors, and preferentially depolarizes the terminals of such afferents, and correspondingly, the other system is activated by and operates on the slowly adapting units. In both PAD systems the size of the depolarization is graded, depending on the stimulus strength.

Furthermore, it was shown by SCHMIDT et al. (1967b) and JÄNIG et al. (1968b) that the PAD of mechanoreceptor afferents of the hind limb is organized in a "surround" fashion. As seen in Fig. 29 the mechanical stimuli exerted their greatest depolarizing influence onto the afferents of those receptors which were nearest to the point of the stimulus application, and the PAD decreased when the conditioning stimulus was moved away from the receptive field of the fibre under study.

Dorsal root potentials evoked by mechanical stimuli were already seen by Barron and Matthews (1938) in the frog and in the cat, and by Fuortes (1951) in the frog. They have also been described in the cat by Mendell and

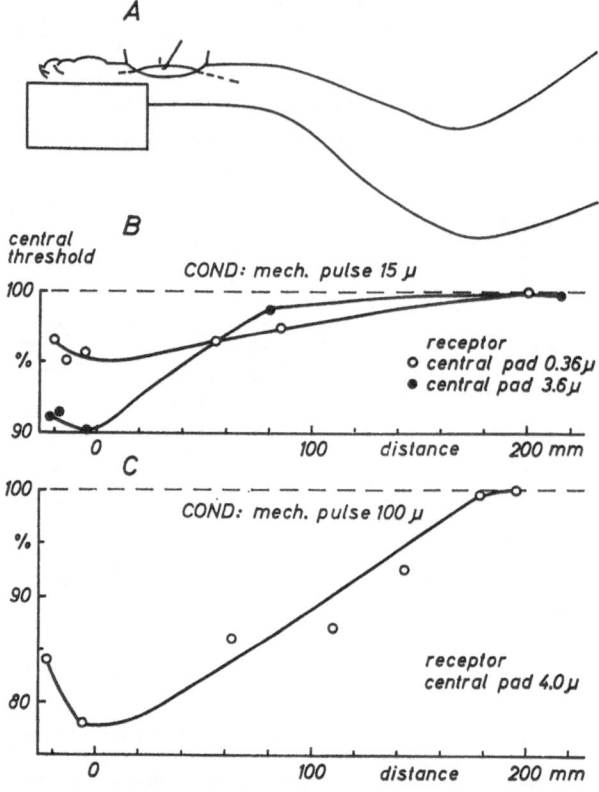

Fig. 29A–C. Topographic organization of PAD of mechanoreceptor afferents. In B the central excitability changes of single mechanoreceptor afferents from the central food pad were measured 50 ms (open circles) and 30 ms (points) after a mechanical stimulus of 15 μ indentation and 4 ms duration applied to various positions along the leg. These distances are given on the abscissa and can be related to the limb in A whic his drawn to scale. The afferent fibres came from touch receptors having mechanical thresholds as indicated in microns. C is from another receptor. The conditioning stimulus produced a 100 μ skin indentation (Schmidt, Senges and Zimmermann, 1967b)

Wall (1964), Vycklický and Tabin (1966), Schmidt, Trautwein and Zimmermann (1966) and Jänig et al. (1967). It is now appreciated that these DRP were evoked by afferent activity in various types of mechanoreceptor afferent fibres, and that they reflected PAD mainly in the Group II myelinated cutaneous afferents of mechanoreceptors.

Painful Stimuli. Intense radiant heat applied to the plantar pad of spinal cats produced PAD of cutaneous fibres (and also, to a lesser extent, of Ib muscle afferents), whether or not the posterior tibial nerve was cooled to block the activity in cutaneous Group II fibres (VYCKLICKÝ et al., 1969; BURKE et al., 1971). WHITEHORN and BURGESS (personal communication) also observed that damaging skin stimuli, such as radiant heat and the application of noxious clips, depolarized cutaneous mechanoreceptor afferents. It was suggested by both groups of authors that these PAD resulted from the afferent activity mainly in small myelinated (A delta) and unmyelinated (C) cutaneous afferent fibres.

3.2.3 Presynaptic Inhibition during PAD of Cutaneous Afferent Fibres

In the motoneurone it has clearly been demonstrated that during PAD of Ia terminals the monosynaptic EPSP is diminished without any other detectable change of the postsynaptic membrane (cf. chapter 1.2). Correspondingly the inhibitory effects of PAD of cutaneous afferent terminals should manifest themselves in a similar reduction of the monosynaptic EPSPs evoked by these terminals in the appropriate segmental interneurones. Because of the small size of these interneurones very few observations of this type have been reported. An example is shown in Fig. 30. In all other cases the conclusion that an inhibitory action was due to presynaptic depolarization was based on the close parallelism observed between the PAD and the inhibition in regard to their various properties such as time course, mode of generation, operational relationships, pharmacology, and so forth.

In addition to the EPSP depression just described, many of the well known inhibitory actions of cutaneous afferent volleys have been attributed in whole or in part to the PAD produced by these volleys in cutaneous afferent terminals. These include: the flexor reflex inhibition produced by ipsilateral (ECCLES, KOSTYUK and SCHMIDT, 1962b; SCHMIDT and WILLIS, 1963b) as well as by contralateral afferent volleys (ECCLES, HOLMQVIST and VOORHOEVE, 1964a, b); the inhibition of the monosynaptic and polysynaptic discharges into the ipsilateral dorsolateral funiculus (ECCLES, KOSTYUK and SCHMIDT, 1962b); the inhibition of dorsal root potentials and dorsal root reflexes (ECCLES, KOSTYUK and SCHMIDT, 1962b; SCHMIDT and WILLIS, 1963b); the inhibition of the discharges of cuneate neurones into the medial lemniscus and the depression of cuneate P-waves (ANDERSEN, ECCLES, OSHIMA and SCHMIDT, 1964; CARLI et al., 1966; CESA-BIANCHI et al., 1968); and the inhibition of discharges in second order neurones of trigeminal nuclei (DARIAN-SMITH, 1965; ROWE and CARMODY, 1970). Furthermore, it has been generally assumed that the PAD in cutaneous afferents produced by other inputs, for example high threshold muscle afferents (FRA-fibres) or supraspinal structures, is always accompanied by a depression of the excitatory actions of the depolarized terminals.

The parallelism between PAD and presynaptic inhibition has so universally been accepted that many authors use both expressions synonymously. Even if it had been established beyond reasonable doubt that PAD always reflects presynaptic inhibition and that presynaptic inhibition is always accompanied by PAD, this synonymous usage should be avoided to prevent misunderstandings. Since at present the rôle of PAD in presynaptic inhibition relative to

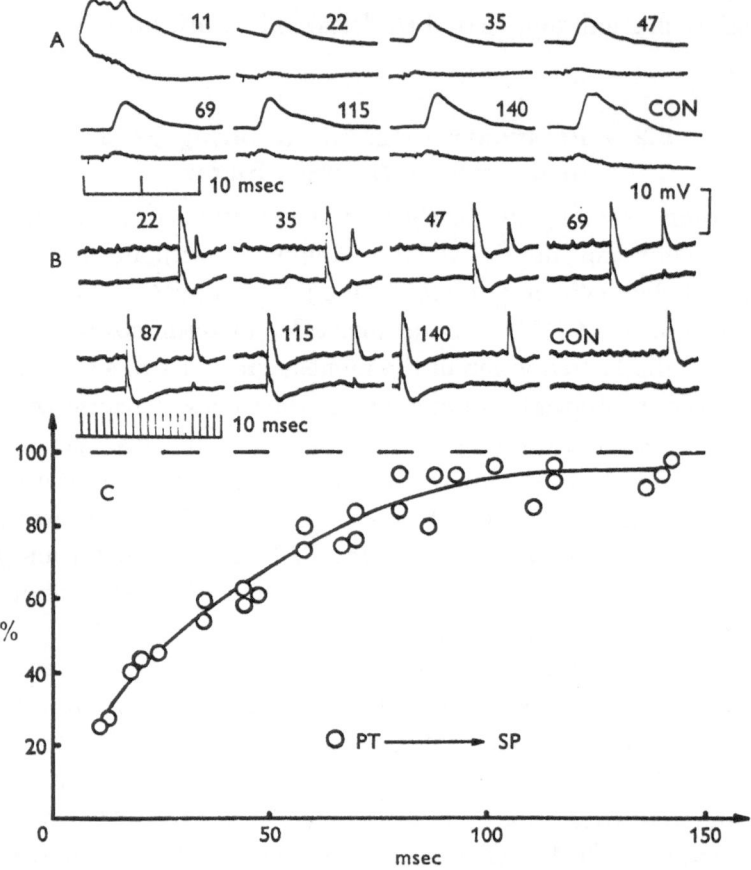

Fig. 30 A–C. Presynaptic inhibition of EPSP of a second order cutaneous interneurone. B shows in the upper trace intracellularly recorded EPSP that were evoked monosynaptically by a single volley in SP nerve and conditioned by another single volley in PT nerve. In A are faster recordings of the EPSP evoked by the testing volley. The numbers in A and B indicate the intervals (ms) between the two volleys. The lower traces in A and B are the cord dorsum surface potentials at L_7 segmental level. In A and B CON is the control EPSP. In C the size of the conditioned EPSP is plotted against the testing intervals for the series partly illustrated in A and B (Eccles, Kostyuk and Schmidt, 1962b)

other factors, such as impedance changes at the terminal regions, is far from being clarified (cf. chapters 1.2 and 2.3.2) it appears even more important not to confuse the terminal depolarization with the accompanying inhibitory process.

3.3 PAD and Presynaptic Inhibition in the Dorsal Column Nuclei

The cuneate and gracile nuclei are first synaptic relays receiving ipsilateral tactile and kinesthetic inputs from forelimb and hindlimb body regions respectively. The primary afferent fibres of the dorsal columns end on relay cells (labelled R in Fig. 31) which project to the contralateral ventrobasal nuclei of the thalamus. In addition the afferent terminals make excitatory synaptic contacts with interneurones which project inside the dorsal column nuclei (for a comprehensive review see NORTON, 1969). THERMAN (1941) first reported that cutaneous volleys passing up the dorsal columns produced on the surface of these nuclei N- and P-waves resembling those produced at the segmental level by the same afferent volleys, and, furthermore, that brief repetitive stimulation of the contralateral sensorimotor cortex evoked a P-wave having a similar time course. It could be expected that the P-waves on the surface of the dorsal column nuclei, just as those at the segmental level, reflected a depolarization of the primary afferent terminals.

The subsequent investigations on the origin of the P-waves evoked by cutaneous volleys or by cortical stimulation clearly showed that the PAD of the dorsal column cutaneous terminals resembled in every respect that seen at the segmental level (WALL, 1958; ANDERSEN, ECCLES and SCHMIDT, 1962; ANDERSEN, ECCLES, SCHMIDT and YOKOTA, 1964a, b; SCHMIDT, VOGEL and ZIMMERMANN, 1966; ANDERSEN, ETHOLM and GORDON, 1967; JABBUR and BANNA, 1968, 1970; DAVIDSON and RYDER, 1969; FELIX and WIESENDANGER, 1970). An example of the intracellularly recorded PAD in the cuneate nucleus has already been given in Fig. 4. An analysis of the properties of cuneate relay cells and interneurones, and of their interconnections led to the diagram shown in Fig. 31 A (ANDERSEN, ECCLES, SCHMIDT and YOKOTA, 1964c). These results and those obtained in a study of the mechanism of synaptic transmission in the cuneate nucleus (ANDERSEN, ECCLES, OSHIMA and SCHMIDT, 1964) led to the conclusion that transmission through the cuneate nucleus can be modified by both presynaptic and postsynaptic inhibitory actions. In Fig. 31 the interneurones on the pre- and postsynaptic inhibitory pathways are labelled P and I respectively. It is seen that somatic activity will excite both pathways, whereas cortical stimulation mainly gives a presynaptic inhibitory action. Comparable results were obtained in a study of the transmission of Group I afferent fibres from forelimb muscles through the cuneate nucleus (ROSÉN, 1969). Pre- and postsynaptic inhibitory phenomena in the dorsal column nuclei can also be observed in unanaesthetized, unrestrained animals (CARLI et al., 1966).

Pharmacological studies also yielded results similar to those obtained in the lumbar cord (cf. chapter 2.5). BOYD et al. (1966) tested the effects of convulsant drugs on the recovery cycles of the cuneate nucleus. They came to the conclusion, schematically shown in Fig. 31 B, that the presynaptic inhibitory pathway is blocked by picrotoxin and pentylenetetrazol, and that there are at least two postsynaptic inhibitory pathways, the shorter one being blocked by strychnine, the other one by strychnine plus mephenesin. BANNA and JABBUR (1969) arrived at similar conclusions when testing the effects of picrotoxin, strychnine and pentobarbitone on the increase in excitability of cuneate presynaptic terminals produced by cortical or cutaneous volleys.

The cortically evoked P-waves differ from those evoked from the cutaneous nerves not only in their remarkable temporal facilitation but also in their slower time course. The neural mechanisms responsible for the temporal

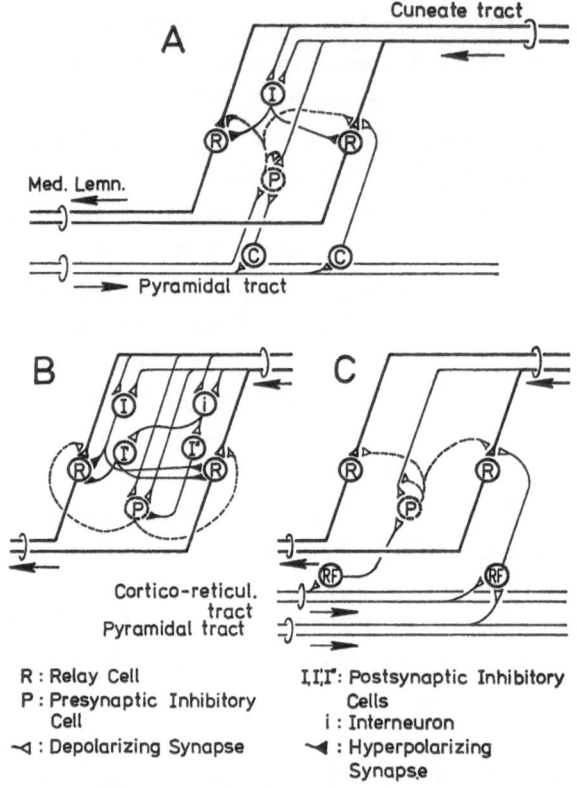

Fig. 31 A–C. Pathways for pre- and postsynaptic inhibition in the dorsal column nuclei. A Schematic diagram illustrating the suggested connections between the primary afferent collaterals, the fibres of the pyramidal tract, and the relay neurones as suggested by ANDERSEN et al. (1964c). B Modification proposed by BOYD, MERITT and GARDNER (1966) of the postsynaptic inhibitory pathways based on pharmacological evidence. C Modification proposed by WIESENDANGER (1969) of the descending presynaptic inhibitory pathways. RF, reticular formation. For detailed discussion see text

facilitation are not located in the cortex, since the facilitation persisted after removal of all the somatosensory cortex and direct stimulation of the pyramidal tract fibres in the underlying white matter (ANDERSEN, ECCLES, SCHMIDT and YOKOTA, 1964a). Hence the neural mechanism for the summation and delayed time course must be situated at a lower level. Circumstantial evidence points to the reticular formation: pyramidotomy at the trapezoid level failed to abolish the cortically induced inhibition at the dorsum column nuclei (JABBUR and TOWE, 1961; LEVITT et al., 1964), and stimulation of the reticular formation inhibited test responses from the dorsal column nuclei (HERNÁNDEZ-PEÓN and

HAGBARTH, 1955; GUZMAN-FLORES et al., 1962; CHAMBERS et al., 1963), and induced PAD of dorsal column primary terminals indiscernible from that evoked from the cortex (CESABIANCHI et al., 1968). Therefore, as shown in Fig. 31 C, the latter authors and WIESENDANGER (1969) proposed that the presynaptic inhibitory interneurones (P-cells, Fig. 31 A, C) not only receive direct excitatory synaptic contacts from pyramidal axon collaterals, but that these neurones also can be activated from the pyramidal tract via the reticular formation.

3.4 PAD and Presynaptic Inhibition of Trigeminal Afferent Fibres

DARIAN-SMITH (1965) first described PAD of trigeminal afferents and a concurrent depression of second-order neurones following electrical and tactile stimulation of the skin at and around the receptive field under observation. He concluded that at least part of the surround inhibition is of presynaptic origin, and this conclusion has been accepted by subsequent investigators (BALDISSERA et al., 1967; STEWART et al., 1967; ROWE, 1970; ROWE and CARMODY, 1970). The general features of peripherally evoked PAD of trigeminal afferents closely resemble that seen in the spinal cord and, particularly, in the dorsal column nuclei. In addition, evidence has been presented that interactions between the various trigeminal nuclei can be exerted via variations of the level of PAD of trigeminal afferents (SCIBETTA and KING, 1969).

The corticofugal inhibitory effects on synaptic transmission in the trigeminal nuclei (HERNANDEZ-PEÓN and HAGBARTH, 1955) parallel in every respect the PAD of trigeminal afferents (DARIAN-SMITH and YOKOTA, 1966a, b; HAMMER et al., 1966; WIESENDANGER et al., 1967a, b; STEWART et al., 1967; SHENDE and KING, 1967; HEPP-REYMOND and WIESENDANGER, 1969; SAUERLAND and MIZUNO, 1969), whereas no signs of postsynaptic inhibition have been detected (WIESENDANGER and FELIX, 1969). Stimulation of the brain stem gives similar effects (BALDISSERA et al., 1967). It has been inferred from these findings that the inhibitory corticofugal control is largely presynaptic in nature and that the pathways are similar to or even identical with those leading to the afferent terminals in the dorsal column nuclei (cf. Fig. 31). The PAD of trigeminal afferents during desynchronized sleep (BALDISSERA, BROGGI and MANCIA, 1966) also has its analogue in the cuneate nucleus (CARLI et al., 1966) and the spinal cord (BALDISSERA, CESA-BIANCHI and MANCIA, 1966).

Recent physiological and anatomical (DUBNER et al., 1969) evidence suggests that the corticofugal fibres leading to the trigeminal brain stem nuclei, while producing PAD of trigeminal primary afferent fibres, are themselves subjected to presynaptic depolarization by trigeminal nerve stimulation. So far, presynaptic depolarization of axon terminals not stemming from primary afferents has been reported only in the somaesthetic pathways (chapter 3.6) and in the visual system (chapter 3.7).

3.5 PAD and Presynaptic Inhibition of Visceral Afferents

Stimulation of splanchnic afferents produces DRPs in the lumbar cord of the cat (Duda et al., 1966; Kostyuk and Preobrazhensky, 1966; Kostyuk, 1968; Hancock, Willis and Harrison, 1970). Visceral afferents of the sympathetic chain and somatic afferent fibres converge in the lumbar spinal cord onto common interneurones in the dorsal horn (Selzer and Spencer, 1969a), and these fibre groups exert a reciprocal PAD on each other (Selzer

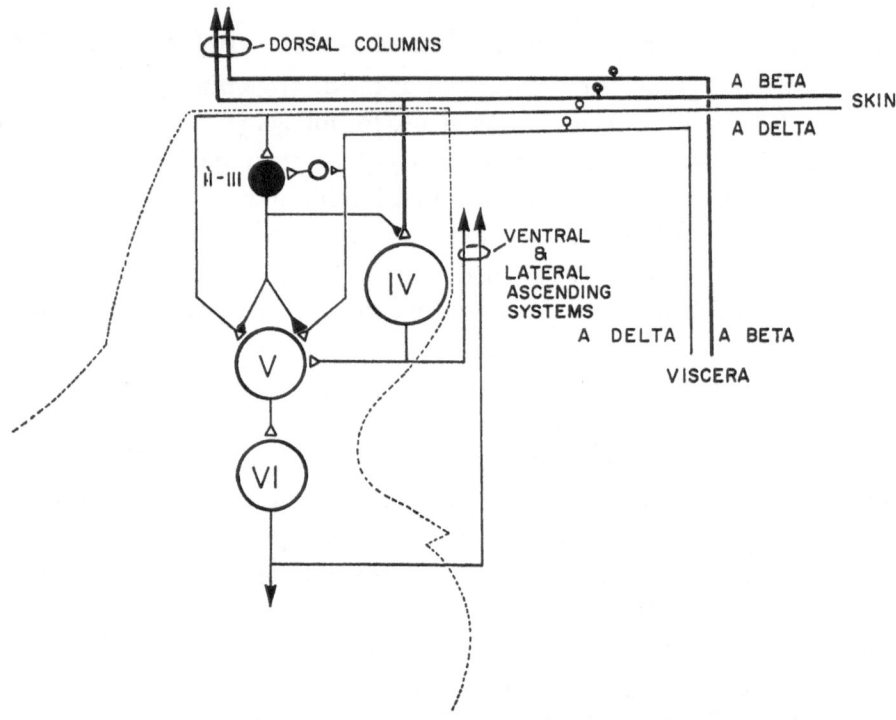

Fig. 32. Pathways giving reciprocal presynaptic inhibition of cutaneous and visceral afferents in the spinal cord. The Roman numerals indicate the dorsal horn layers according to Rexed (1952, 1954). Presynaptic interactions between and within different cutaneous fibre groups (cf. Fig. 28), and possible cutaneous delta fibre connections to lamina IV cells have been omitted. See text for further details (Selzer and Spencer, 1969b)

and Spencer 1969b; Hancock et al., 1970). The PAD is paralleled by an inhibition of the actions of the depolarized fibres on dorsal horn interneurones, indicating that these two afferent systems exert some reciprocal presynaptic inhibition. Selzer and Spencer (1969b), on the basis of their own and of other available data, constructed the diagram shown in Fig. 32 of the interactions of the visceral and cutaneous afferent systems. They concluded that the convergence of visceral and cutaneous afferent pathways provides a mechanism by which the reference of visceral pain to the skin might be explained, and that the pronounced inhibition of visceral afferent actions by conditioning cutaneous afferent volleys may underlie the clinical phenomenon of inhibition of visceral pain by cutaneous "counter-irritation".

The afferent fibres of the superior laryngeal nerve innervate mucosal receptors of the larynx, and their presynaptic terminals end in the solitary tract nucleus. In a series of publications RUDOMIN (1966, 1967a, b, 1968) showed that the presynaptic terminals of these fibres can be depolarized by stimulation of low threshold vagal and aortic afferents (presumably from pulmonary stretch receptors and circulatory pressor receptors, respectively), and also by single and repetitive stimulation of the solitary tract nucleus. No PAD of vagal or aortic nerve terminals in the solitary tract nucleus was detected under the same conditions. In agreement with these latter findings, FRANKSTEIN and SERGEEVA (1967) reported that stimulation of the cervical vagus produced a small or no P-wave from the surface of the medulla at the site of projection of the tractus solitarius. RUDOMIN as well as FRANKSTEIN and SERGEEVA concluded from these findings that vagal and depressor afferents are not subjected to presynaptic control. However, WEISS and CRILL (1969) clearly showed a PAD of carotid sinus nerve afferents in the solitary tract following stimulation in the posterior hypothalamus. It appears from these results that the vagal visceral afferents do not receive appreciable PAD from the peripheral inputs tested so far, but that the supramedullary control of cardiovascular and pulmonary reflexes may at least partly be exerted via presynaptic inhibition of the input from the respective peripheral receptors. The absence of PAD in visceral afferents following local stimulation of the solitary tract nucleus (RUDOMIN, 1968) perhaps merely indicates that the interneurones mediating the descending PAD are situated outside this nucleus.

3.6 Presynaptic Inhibition of Axon Terminals of Second-Order Cells in Somato-Sensory Pathways

The results reviewed in chapters 3.1–3.5 permit the assumption that in the vertebrate central nervous system all types of myelinated somesthetic primary afferent fibres are subject to PAD and thus to a presynaptic inhibitory control of their excitatory actions. Very little is known about presynaptic inhibition of axon terminals stemming from other than primary afferent axons. It is generally felt among neurophysiologists that presynaptic inhibition is restricted mainly to primary afferent terminals, but sufficient evidence to substantiate this opinion is lacking. Definite indications of presynaptic inhibition of axon terminals of central neurones have been seen in the trigeminal nuclei (chapter 3.4; DUBNER et al., 1969) and in somesthetic relay paths (see below). In addition, in the lateral geniculate nucleus, morphological and physiological evidence for presynaptic inhibition of optic tract endings and of other axon terminals has been reported (chapters 2.2.3 and 3.7).

The axons of the second-order cells of the dorsal column nuclei project to the thalamus through the contralateral medial lemniscus. Excitability testing of the axon terminals of cuneate tract fibres in the ventro-basal complex of the thalamus before and after conditioning stimuli to forelimb nerves revealed a depolarization with a time course characteristic for PAD of spinal afferents, and it was concluded that this depolarization reflected presynaptic inhibition of lemniscal terminals which added to the powerful postsynaptic inhibition of the thalamo-cortical relay cells observed under these conditions (ANDERSEN, BROOKS, ECCLES and SEARS, 1964). There were no clear indications of presynaptic inhibition induced by cortical stimulation. The pathways of post-

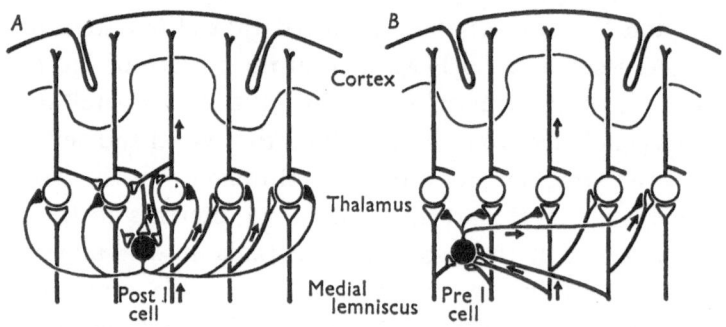

Fig. 33A and B. Pathways for postsynaptic (A) and presynaptic inhibition (B) in the ventro-basal complex of the thalamus. In A the axon collaterals of thalamocortical relay cells (TRC) are seen to excite both TRC and the postsynaptic inhibitory interneurone (Post I cell). In B branches of lemniscal fibres excite the presynaptic inhibitory interneurone (Pre I cell), which is widely distributed to the synaptic knobs of lemniscal fibres (ANDERSEN, ECCLES and SEARS, 1964)

synaptic and presynaptic inhibition (Fig. 33) have been further analyzed by ANDERSEN, ECCLES and SEARS (1964a). By employing lemniscal and cortical stimulation it was shown that ventro-basal cells could be identified either as thalamo-cortical relay cells or as cells which qualified as being postsynaptic or presynaptic inhibitory interneurones.

The spinocervical tract originates from second-order spinal dorsal horn neurones and terminates on cells of the lateral cervical nucleus. (The latter cells project to the thalamus through the contralateral medial lemniscus). In a short communication FEDINA et al. (1968) presented evidence that the cells of the lateral cervical nucleus are subject to both pre- and postsynaptic inhibitory actions. Two indications for presynaptic inhibition were found: electrical stimulation of cutaneous nerves and of the skin of the cat's forepaw increased the excitability of spinal cervical tract axons; at the same time the monosynaptically evoked EPSPs in cells of the lateral cervical nucleus were reduced without any change of the membrane potential of the cells. No evidence for a "surround" pattern of inhibition was found. Both pre- and postsynaptic inhibitory actions could be evoked from large parts of the body.

3.7 Presynaptic Inhibition in the Visual System

A number of investigations have drawn attention to the fact that subcortical and cortical influences on synaptic transmission in the lateral geniculate body, LGB, are accompanied and often paralleled by a depolarization of the optic tract terminals in that nucleus. For instance, stimulation of the mesencephalic reticular formation produced an excitability increase of optic fibres with the same latency and time course as the PAD of spinal afferents induced by brain stem or cortical stimulation (ANGEL et al., 1965, 1966). Intracellular recording from optic tract fibres produced conclusive evidence that these excitability changes were due to a transmembrane depolarization generated by an active mechanism (KAHN et al., 1967). Stimulation of the vestibular system (MARCHIAFAVA and POMPEIANO, 1966a, b) or of various areas of the visual and neighbouring cortex produced comparable effects (IWAMA et al., 1965; SUZUKI and KATO, 1965; ANGEL, MAGNI and STRATA, 1967). The presynaptic changes were blocked by picrotoxin, but not by strychnine, and they were prolonged by barbiturates (PECCI-SAAVEDRA et al., 1966). In chronic unrestrained cats large excitability changes of the optic tract terminals in the LGB occurred during the rapid eye movements of desynchronized sleep (BIZZI, 1966; IWAMA et al. 1966); similar changes were induced during ocular movements of the alert cat (KAWAMURA and MARCHIAFAVA, 1966), and a particularly striking excitability increase was measured during epileptic EEG activity evoked by stimulation of the sensorimotor cortex (ANGEL and STRATA, 1967).

This considerable body of physiological evidence in favour of a presynaptic inhibitory mechanism acting on the terminals of optic tract fibres is supplemented by several histological reports on axo-axonic synapses in the complex glomerular structures of the LGB (see chapter 3.2.3). It has been generally assumed, therefore, that the preterminal depolarization of the optic tract fibres reflects an inhibition of the synaptic efficacy of the optic tract terminals.

The functional significance of this presumed presynaptic inhibitory control of transmission of information through the LGB is at present largely a matter of speculation. This is not surprising since the synaptic organization and the transfer of retinal and nonretinal input into the LGB is far from being fully understood (see the recent review by CREUTZFELDT and SAKMANN, 1969).

4. Summarizing and Concluding Remarks

4.1 Mechanism of Primary Afferent Depolarization and Presynaptic Inhibition

Our present state of knowledge of the mechanisms producing presynaptic depolarization of primary afferent fibres (primary afferent depolarization,

PAD), and of the inhibitory effects induced thereby, may be summarized as follows (numbers in brackets refer to the appropriate parts of the review):

1. Primary afferent fibres, and possibly the axons of several other types of neurones, can be subjected to a prolonged depolarization which can be recorded with a variety of electrophysiological techniques (chapter 2.1).

2. Physiological evidence indicates that this depolarization is exerted at or near the terminals of the axon (chapter 2.3.1).

3. The presynaptic depolarization blocks or decreases the action potential propagating towards the axon terminals.

4. As a consequence, a reduced amount of the transmitter is released by the presynaptic action potential (chapter 2.3.3). This inhibitory process has been called presynaptic inhibition.

5. All phenomena are in good agreement with the concept that the presynaptic depolarization is due to the activation of chemically mediated synapses situated at or near the preterminal endings of the axons subjected to presynaptic depolarization (chapters 1.2 and 2.3.2). Anatomical evidence supports this concept originally derived from physiological findings (chapter 2.2).

6. The axo-axonal synapses can be activated through reflex pathways consisting of two to several interneurones (chapter 2.4).

7. The pharmacology of presynaptic inhibition differs remarkably from that of postsynaptic inhibition.

The following aspects of this concept urgently need further clarification: first of all, future experiments have to test further the hypothesis of the axo-axonal mechanism of presynaptic depolarization. Secondly, the changes which the presynaptic spike undergoes during PAD are not known in any detail. And thirdly, so far only relatively indirect evidence has been brought forward in favour of the postulated steep relationship between presynaptic spike size and transmitter release.

4.2 Functional Significance of Presynaptic Inhibition in the Spinal Cord

It has to be appreciated that the evaluation of the functional rôle of presynaptic inhibition in the vertebrate central nervous system is just as impossible a task as that of describing the functional rôle of postsynaptic inhibition. An attempt can be made, however, to assess the significance of presynaptic inhibition of primary afferent fibres of the spinal cord. The considerable work done on this aspect of presynaptic inhibition can be summarized as follows:

1. The most remarkable feature of segmental PAD is its negative feedback character. The various muscle and cutaneous efferents preferentially activate the PAD reflex paths leading to the afferent terminals of fibres of their own sensory modality or quality. The advantages of this type of negative feedback

include both a central input adjustment to the stimulus intensity level, and the automatic suppression of trivial inputs so that the central mechanism is cleared in readiness for significant inputs.

2. Within the Group I afferent fibres from the muscles of a given limb those coming from flexor muscles are generally more powerful than those coming from extensors, and the Ib fibres are much more powerful than the Ia fibres. The significance of these findings for the reflex control of movement and posture (LUNDBERG, 1966; HOUK and HENNEMAN, 1967) is an open question, much as is the functional significance of the postsynaptic Renshaw inhibition (chapter 3.1).

3. For the cutaneous afferents coming from mechano-receptors, it has been demonstrated that the negative feedback is organized in a "surround" fashion, thus creating a system of lateral or afferent inhibition at the earliest possible level of the somatic afferent system. This arrangement gives, for instance, the possibility of restoration of spatial contrast in the stimulus pattern and it facilitates the localization of the stimulus (chapter 3.2.2).

4. Descending presynaptic inhibitory influences from various parts of the brain stem and the cortex are able to increase or decrease selectively the somaesthetic and proprioceptive input channels thus providing a mechanism of "focussing the attention" on important messages (chapters 3.1, 3.2.1, 3.3). The sensitivity of the channels can be decreased through PAD, whereas it can be increased by a removal or suppression of presynaptic inhibition (disinhibition) coming from other sources, which is recordable, for instance, as a presynaptic excitability increase or as a "positive" DRP, see chapter 3.1.2. Habituation and sensitization are not paralleled by long-lasting PAD or PAH (GROVES et al., 1970).

The results summarized in 1–4 above have been contributed by a number of different laboratories. It is remarkable that the research reports as far as they overlap agree with each other to an extent rarely found in biological research. Some controversy, however, has recently developed on the contribution of high threshold afferent fibres (Group III and IV fibres) to the PAD of myelinated afferents. Following electrical stimulation of cutaneous Group III and IV fibres two contrary findings have been reported: presynaptic hyperpolarizations (positive DRP, MENDELL and WALL, 1964; DAWSON et al., 1970); and presynaptic depolarizations (FRANZ and IGGO, 1969; ZIMMERMANN, 1968 b; JÄNIG and ZIMMERMANN, 1971). In addition MENDELL (1970) claimed that stimulation of small diameter muscle afferents produced positive DRPs. These diverse results aroused considerable interest because it had been proposed that the disinhibition reflected in the positive DRPs, might play a rôle in the perception of noxious stimuli (gate-control theory of pain, MELZACK and WALL, 1965). Noxious stimuli to the skin, however, only led to primary afferent depolarization, and it was concluded that these results do not support the proposal of a gating mechanism operating presynaptically during painful stimuli (Vycklicky et al., 1969; Burke et al., 1971; Whitehorn and Burgess, personal communication). To resolve the discrepancies it appears necessary not only to compare carefully and to control the experimental conditions but also to activate the various C-fibre modalities by adequate stimulation and to use recording methods which precisely determine the types of afferent fibres receiving the depolarization.

Acknowledgement: The author is indebted to Drs. G. Allen, J. Dudel, J. C. Eccles, M. Rowe and M. Zimmermann for valuable comments and suggestions and for reading the manuscript.

List of Abbreviations in Alphabetical Order

Afferent Fibres

a) Muscle Afferents

Group Ia	from primary muscle spindle endings
Group Ib	from Golgi tendon organs
Group II	from secondary muscle spindle endings
Group III	high threshold myelinated afferents
Group IV	= C-fibres, unmyelinated

b) Cutaneous Afferents

Group II	low threshold myelinated afferents
Group III	high threshold myelinated afferents
Group IV	= C-fibres, unmyelinated

D-cell	Dorsal horn interneurone on PAD reflex pathway
DCR	Dorsal column reflex
DRP	Dorsal root potential
DR-DRP	Dorsal root potential evoked in the amphibian spinal cord by stimulation of dorsal roots
DRR	Dorsal root reflex
EPSP	Excitatory postsynaptic potential
IPSP	Inhibitory postsynaptic potential
LGB	Lateral geniculate body
N-wave	Negative potential wave
PAD	Primary afferent depolarization
PAH	Primary afferent hyperpolarization
PC-receptor	Rapidly adapting mechanoreceptor (Pacinian corpuscle)

Peripheral Nerves

FDHL	flexor digitorum plus hallucis longus
GS	gastrocnemius plus soleus
M	median
PBST	posterior biceps plus semitendinosus
PDP	peroneal plus deep peroneal (flexor branches only)
PL	plantaris
SU	suralis
PT	posterior tibial
SMAB	semimembranosus plus anterior biceps
SP	superficial peroneal
SR	superficial radial
U	ulnaris

P-wave   positive potential wave
RA-receptor   rapidly adapting mechanoreceptor of hairless skin
SA-receptor   slowly adapting mechanoreceptor of hairless skin
SG-cell   Substantia gelatinosa cell
VR-DRP   dorsal root potential evoked in the amphibian spinal cord by stimulation
              of ventral roots

References

ANDÉN, N. E., JUKES, M. G. M., LUNDBERG, A.: Spinal reflexes and monoamine liberation. Nature (Lond.) **202**, 1222–1223 (1964).

— — — The effect of Dopa on the spinal cord. 2. A pharmacological analysis. Acta physiol. scand. **67**, 387–397 (1966).

— — — VYKLICKÝ, L.: A new spinal flexor reflex. Nature (Lond.) **202**, 1344–1345 (1964).

— — — — The effect of Dopa on the spinal cord. 1. Influence on transmission from primary afferents. Acta physiol. scand. **67**, 373–380 (1966a).

— — — — The effect of Dopa on the spinal cord. 3. Depolarization evoked in the central terminals of ipsilateral I a afferents by volleys in the flexor reflex afferents. Acta physiol. scand. **68**, 322–336 (1966b).

— LUNDBERG, A., ROSENGREN, E., VIKLICKÝ, L.: The effect of Dopa on spinal reflexes from the FRA (flexor reflex afferents). Experientia (Basel) **19**, 654–655 (1963).

ANDERSEN, P., BROOKS, CH. McC., ECCLES J. C., SEARS, T. A.: The ventro-basal nucleus of the thalamus: potential fields, synaptic transmission and excitability of both presynaptic and post-synaptic components. J. Physiol. (Lond.) **174**, 348–369 (1964).

— ECCLES, J. C., OSHIMA, T., SCHMIDT, R. F.: Mechanisms of synaptic transmission in the cuneate nucleus. J. Neurophysiol. **27**, 1096–1116 (1964).

— — SCHMIDT, R. F.: Presynaptic inhibition in the cuneate nucleus. Nature (Lond.) **194**, 741–743 (1962).

— — — YOKOTA, T.: Slow potential waves produced in the cuneate nucleus by cutaneous volleys and by cortical stimulation. J. Neurophysiol. **27**, 78–91 (1964a).

— — — — Depolarization of presynaptic fibers in the cuneate nucleus. J. Neurophysiol. **27**, 92–106 (1964b).

— — — — Identification of relay cells and interneurons in the cuneate nucleus. J. Neurophysiol. **27**, 1080–1095 (1964c).

— — SEARS, T. A.: Presynaptic inhibitory action of cerebral cortex on the spinal cord. Nature (Lond.) **194**, 740–743 (1962).

— — — Cortically evoked depolarization of primary afferent fibers in the spinal cord. J. Neurophysiol. **27**, 63–77 (1964a).

— — — The ventro-basal complex of the thalamus: types of cells, their responses and their functional organization. J. Physiol. (Lond.) **174**, 370–399 (1964b).

— ETHOLM, B., GORDON, G.: Presynaptic depolarization of dorsal column fibres by adequate stimulation. J. Physiol. (Lond.) **194**, 83P–84P (1967).

ANGEL, F., MAGNI, P., STRATA, P.: Evidence from pre-synaptic inhibition in the lateral geniculate body. Nature (Lond.) **208**, 495–496 (1965).

— — — Excitability of intra-geniculate optic tract fibres after reticular stimulation in the midpontine pretrigeminal cat. Arch. ital. Biol. **103**, 668–693 (1966).

— — — The excitability of optic nerve terminals in the lateral geniculate nucleus after stimulation of visual cortex. Arch. ital. Biol. **105**, 104–117 (1967).

— STRATA, P.: Relationship between cortical activity and the excitability of optic nerve terminals in the lateral geniculate body. Brain Res. **5**, 501–503 (1967).

AUSTIN, G. M., McCOUCH, G. P.: Presynaptic component of intermediary cord potential. J. Neurophysiol. **18**, 441–451 (1955).

BALDISSERA, F., BROGGI, G.: An analysis of potential changes in the spinal cord during desynchronized sleep. Brain Res. **6**, 706–715 (1967).

Baldissera, F., Broggi, G., Mancia, M.: Presynaptic inhibition of trigeminal afferent fibres during the rapid eye movements of desynchronized sleep. Experientia (Basel) **22**, 754–755 (1966).
— — — Depolarization of trigeminal afferents induced by stimulation of brain-stem and peripheral nerves. Exp. Brain Res. **4**, 1–17 (1967).
— Cesa-Bianchi, M. G., Mancia, M.: Phasic events indicating presynaptic inhibition of primary afferents to the spinal cord during desynchronized sleep. J. Neurophysiol. **29**, 871–887 (1966).
Banna, N. R., Hazbun, J.: Analysis of the convulsant action of pentylenetetrazol. Experientia (Basel) **25**, 382–383 (1969).
— Jabbur, S. J.: Antagonism of presynaptic inhibition in the cuneate nucleus by picrotoxin. Nature (Lond.) **217**, 83–84 (1968).
— — Pharmacological studies on inhibition in the cuneate nucleus of the cat. Int. J. Neuropharmacol. **8**, 299–308 (1969).
— — The action of bemegride on presynaptic inhibition. Neuropharmacology **9**, 553–560 (1970).
Barnes, C. D., Pompeiano, O.: Presynaptic inhibition of extensor monosynaptic reflex by Ia afferents from flexors. Brain Res. **18**, 380–383 (1970a).
— — Effects of muscle vibration on the pre- and post-synaptic components of the extensor monosynaptic reflex. Brain Res. **18**, 384–387 (1970b).
— — Inhibition of monosynaptic extensor reflex attributable to presynaptic depolarization of the group Ia afferent fibers produced by vibration of flexor muscle. Arch. ital. Biol. **108**, 233–258(1970c).
— — Presynaptic and postsynaptic effects in the monosynaptic reflex pathway to extensor motoneurons following vibration of synergic muscles. Arch. ital. Biol. **108**, 259–294 (1970d).
Barron, D. H.: Central course of 'recurrent' sensory discharges. J. Neurophysiol. **3**, 403–406 (1940).
— Matthews, B. H. C.: Intermittent conduction in the spinal cord. J. Physiol. (Lond.) **85**, 73–103 (1935).
— — The interpretation of potential changes in the spinal cord. J. Physiol. (Lond.) **92**, 276–321 (1938).
Benoit, P. R., Mambrini, J.: Modification of transmitter release by ions which prolong the presynaptic action potential. J. Physiol. (Lond.) **210**, 681–696 (1970).
Bernhard, C. G.: The cord dorsum potentials in relation to peripheral source of afferent stimulation. Cold Spr. Harb. Symp. quant. Biol. **17**, 221–232 (1952).
— The spinal cord potentials in leads from the cord dorsum in relation to peripheral source of afferent stimulation. Acta physiol. scand. **29**, Suppl. **106**, 1–29 (1953a).
— Analysis of the spinal cord potentials in leads from the cord dorsum. In: The spinal cord, Ciba Foundation Symp., p. 43–60, ed. G. E. W. Wolstenholme, London 1953b.
— Koll, W.: On the effects of strychnine, asphyxia and dial on the spinal cord potentials. Acta physiol. scand. **29**, Suppl. **106**, 30–41 (1953).
— Widén, L.: On the origin of the negative and positive cord potentials evoked by stimulation of low threshold cutaneous fibres. Acta physiol. scand. **29**, Suppl. **106**, 42–54 (1953).
Bizzi, E.: Changes in the orthodromic and antidromic response of optic tract during the eye movements of sleep. J. Neurophysiol. **29**, 861–870 (1966).
Bloedel, J., Gage, P. W., Llinás, R., Quastel, D. M. J.: Transmitter release at the squid giant synapse in the presence of tetrodotoxin. Nature (Lond.) **212**, 49–50 (1966).
— — — — Transmission across the squid giant synapse in the presence of tetrodotoxin. J. Physiol. (Lond.) **188**, 52–53P (1967).
Bonnet, V., Bremer, F.: Études des potentiels électriques de la moelle épinière faisant suite chez la grenouille spinale à une ou deux volées d'influx centripètes. C. R. Soc. Biol. (Paris) **127**, 806–812 (1938).

BONNET, V., BREMIR, F.: Du mécanisme de l'inhibition centrale. C. R. Soc. Biol. (Paris) **130**, 760–767 (1939).

— — La transmission synaptique dans la substance grise spinale. J. Physiol. Path. gén. **40**, 117A–119A (1948).

— — Les potentiels synaptiques et la transmission nerveuse centrale. Arch. int. Physiol. **60**, 33–93 (1952).

BOYD, E. S., MERITT, D. A., GARDNER, L. C.: The effect of convulsant drugs on transmission through the cuneate nucleus. J. Pharmacol. exp. Ther. **154**, 398–409 (1966).

BRAUN, M., SCHMIDT, R. F.: Potential changes recorded from the frog motor nerve terminal during its activation. Pflügers Arch. ges. Physiol. **287**, 56–80 (1966).

BRAVO, M. C., MOLINA, A. FERNANDEZ DE: Frog's spinal cord potentials generated by stimulation of cutaneous nerves. J. Physiol. (Lond.) **155**, 86–97 (1961).

BREMER, F.: Le tetanos strychnine et le mécanisme de la synchronisation neuronique. Arch. int. Physiol. **51**, 211–260 (1941).

— Le mode d'action de la strychnine à la lumière de travaux récents. Arch. int. Pharmacodyn. **69**, 249–264 (1944).

— Strychnine tetanus of the spinal cord. In: The spinal cord, Giba Found. Symp., ed. G. E. W. Wolstenholme. p. 78–83. London: Churchill 1953.

— BONNET, V.: Contributions à l'étude de la physiologie générale des centres nerveux. II. L'inhibition réflexe. Arch. int. Physiol., **52**, 153–194 (1942).

— — Les potentiels synaptiques et leur interprétation. Arch. Sci. physiol. **3**, 489–520 (1949).

BROOKS, C. McC., ECCLES, J. C., MALCOLM, J. L.: Synaptic potentials of inhibited motoneurones. J. Neurophysiol. **11**, 417–430 (1948).

— FUORTES, M. G. F.: The relation of dorsal and ventral root potentials to reflex activity in mammals. J. Physiol. (Lond.) **116**, 380–394 (1952).

— KOIZUMI, K.: Origin of the dorsal root reflex. J. Neurophysiol. **19**, 61–74 (1956).

— — MALCOLM, J. L.: Effects of changes in temperature on reactions of spinal cord. J. Neurophysiol. **18**, 205–216 (1955).

BURKE, R. E., RUDOMIN, P., VYKLICKÝ, L., ZAJAC, F. E.: Primary afferent depolarization and flexion reflexes produced by radiant heat stimulation of the skin. J. Physiol. (Lond.) **213**, 185—214 (1971).

CALMA, I.: Presynaptic inhibition of the terminals of cutaneous nerve fibres by stimulation of the ventral thalamo-diencephalic region. J. Physiol. (Lond.) **185**, 58P–60P (1966).

— QUAYLE, A. A.: Repetitive antidromic discharges in fast cutaneous nerve fibers. Brain Res. **11**, 268–272 (1968).

CANGIANO, A., COOK, W. A., POMPEIANO, O.: Primary afferent depolarization in the lumbar cord evoked from the fastigial nucleus. Arch. ital. Biol. **107**, 321–340 (1969a).

— — — Cerebellar inhibitory control of the vestibular reflex pathways to primary afferents. Arch. ital. Biol. **107**, 341–364 (1969b).

CARLI, G., DIETE-SPIFF, K., POMPEIANO, O.: Presynaptic and postsynaptic inhibition on transmission of cutaneous afferent volleys through the cuneate nucleus during sleep. Experientia (Basel) **22**, 239–240 (1966).

CARLSSON, C. B.: Sodium and the dorsal root potential. J. Physiol. (Lond.) **172**, 295–304 (1964).

— FALCK, B., FUXE, K., HILLARP, N.: Cellular localization of monoamines in the spinal cord. Acta physiol. scand. **60**, 112–119 (1964).

CARPENTER, D., ENGBERG, I., LUNDBERG, A.: Presynaptic inhibition in the lumbar cord evoked from the brain stem. Experientia (Basel) **18**, 450–451 (1962).

— — — Primary afferent depolarization evoked from the brain stem and the cerebellum. Arch. ital. Biol. **104**, 73–85 (1966).

— LUNDBERG, A., NORRSELL, U.: Effects from the pyramidal tract on primary afferents and on spinal reflex actions to primary afferents. Experientia (Basel) **18**, 337–338 (1962).

CARPENTER, D., ENGBERG, I., LUNDBERG, A.; Primary afferent depolarization evoked from the sensorimotor cortex. Acta physiol. scand. **59**, 126–142 (1963).

CASTILLO, J. DEL, KATZ, B.: Changes in end-plate activity produced by presynaptic polarization. J. Physiol. (Lond.) **124**, 586–604 (1954a).

— — The membrane change produced by the neuromuscular transmitter. J. Physiol. (Lond.) **125**, 546–565 (1954b).

CESA-BIANCHI, M. G., MANCIA, M., SOTGIU, M. L.: Depolarization of afferent fibers to the Goll and Burdach nuclei induced by stimulation of the brain-stem. Exp. Brain Res. **5**, 1–15 (1968).

CHAMBERS, W. W., LIU, CH. N., McCOUCH, G. P.: Inhibition of the dorsal column nuclei. Exp. Neurol. **7**, 13–23 (1963).

CHU, N.-S.: Dorsal root potentials: effects of acoustic and visual stimuli. Brain Res. **18**, 189–191 (1970).

COLONNIER, M., GUILLERY, R. W.: Synaptic organization in the lateral geniculate nucleus of the monkey. Z. Zellforsch. **62**, 333–355 (1964).

CONRADI, S.: On motoneuron synaptology in adult cats. Acta physiol. scand., Suppl. **332**, 1–115 (1969).

COOK, W. A., CANGIANO, A., POMPEIANO, O.: Vestibular influences on primary afferents in the spinal cord. Pflügers Arch. ges. Physiol. **289**, 334–338 (1968).

— — — Dorsal root potentials in the lumbar cord evoked from the vestibular system. Arch. ital. Biol. **107**, 275–295 (1969a).

— — — Vestibular control of transmission in primary afferents to the lumbar spinal cord. Arch. ital. Biol. **107**, 296–320 (1969b).

— NEILSON, D. R., BROOKHART, J. M.: Primary afferent depolarization and mono-synaptic reflex depression following succinylcholine administration. J. Neurophysiol. **28**, 280–311 (1965).

COOMBS, J. S., CURTIS, D. R., LANDGREN, S.: Spinal cord potentials generated by impulses in muscle and cutaneous afferent fibers. J. Neurophysiol. **19**, 452–467 (1956).

CREUTZFELDT, O., SAKMAN, B.: Neurophysiology of vision. Ann. Rev. Physiol. **31**, 499–544 (1969).

CURTIS, D. R.: The pharmacology of central and peripheral inhibition. Pharmacol. Rev. **15**, 333–364 (1963).

— Pharmacology and neurochemistry of mammalian central inhibitory processes. In: Structure and function of inhibitory neuronal mechanisms, ed. C. von Euler, S. Skoglund, U. Söderberg, p. 429–456. Oxford-New York: Pergamon Press 1969.

— The pharmacology of spinal postsynaptic inhibition. Progr. in Brain Res. **31**, 171–189 (1969).

— DUGGAN, A. W., FELIX, D., JOHNSTON, G. A. R.: GABA, bicuculline and central inhibition. Nature (Lond.) **226**, 1222–1224 (1970a).

— — — — Bicuculline and central GABA receptors. Nature (Lond.) **228**, 676–677 (1970b).

— ECCLES, J. C.: Synaptic action during and after repetitive stimulation. J. Physiol. (Lond.) **150**, 374–398 (1960).

— PHILLIS, J. W., WATKINS, J. C.: Actions of amino acids on the isolated hemisected spinal cord of the toad. Brit. J. Pharmacol. **16**, 262–283 (1961).

— RYALL, R. W.: Pharmacological studies upon spinal presynaptic fibres. Exp. Brain Res. **1**, 195–204 (1966).

DAHLSTRÖM, A., FUXE, K.: Evidence for the existence of monoamine neurons in the central nervous system. II. Experimentally induced changes in the intraneuronal amine levels of bulbo-spinal neuron systems. Acta physiol. scand. **64**, Suppl. 247, 1–85 (1965).

DARIAN-SMITH, I.: Presynaptic component in the afferent inhibition observed within trigeminal brain-stem nuclei of the cat. J. Neurophysiol. **28**, 695–709 (1965).

DARIAN-SHMITH, I., YOKOTA, T.: Cortically evoked depolarization of trigeminal cutaneous afferent fibres in the cat. J. Neurophysiol. **29**, 170–184 (1966a).
— — Corticofugal effects on different neuron types within the cat's brain stem activated by tactile stimulation on the face. J. Neurophysiol. **29**, 185–206 (1966b).
DAVIDSON, N., RYDER, C. A.: Interneurone responses in the rat cuneate nucleus. J. Physiol. (Lond.) **204**, 79P (1969).
DAWSON, G. D., MERRILL, E. G., WALL, P. D.: Dorsal root potentials produced by stimulation of fine afferents. Science **167**, 1385–1387 (1970).
DECANDIA, M., GASTEIGER, E. L., MANN, M. D.: Escape of the extensor monosynaptic reflex from presynaptic inhibition. Brain Res. **7**, 317–319 (1968).
— PROVINI, L., TÁBORÍKOVÁ, H.: Excitability changes in the Ia extensor terminals induced by stimulation of agonist afferent fibres. Brain Res. **2**, 402–404 (1966).
— — — Presynaptic inhibition of the monosynaptic reflex following the stimulation of nerves to extensor muscles of the ankle. Exp. Brain Res. **4**, 34–42 (1967).
DECIMA, E. E.: An effect of postsynaptic neurons upon presynaptic terminals. Proc. nat. Acad. Sci. (Wash.) **63**, 58–64 (1969).
— GOLDBERG, L. J.: Time course of excitability changes of primary afferent terminals as determined by motoneuron-presynaptic interaction. Brain Res. **15**, 288–290 (1969).
— — Centrifugal dorsal root discharges induced by motoneurone activation. J. Physiol. (Lond.) **207**, 103–118 (1970).
DEVANANDAN, M. S., ECCLES, R. M., STENHOUSE, D.: Presynaptic inhibition evoked by muscle contraction. J. Physiol. (Lond.) **185**, 471–485 (1966).
— — YOKOTA, T.: Presynaptic inhibition induced by muscle stretch. Nature (Lond.) **204**, 996–998 (1964).
— — — Depolarization of afferent terminals evoked by muscle stretch. J. Physiol. (Lond.) **179**, 417–429 (1965a).
— — — Muscle stretch and the presynaptic inhibition of the group Ia pathway to motoneurones. J. Physiol. (Lond.) **179**, 430–441 (1965b).
— HOLMQVIST, B., YOKOTA, T.: Presynaptic depolarization of group I muscle afferents by contralateral afferent volleys. Acta physiol. scand **63**, 46–54 (1965).
DUBNER, R., SESSLE, B. J., GOBEL, S.: Presynaptic depolarization of corticofugal fibres participating in a feedback loop between trigeminal brain stem nuclei and sensorimotor cortex. Nature (Lond.) **223**, 72–73 (1969).
DUDA, P., KOSTYUK, P. G., PREOBRAZHENSKY, N. N.: Inhibition of synaptic potentials in motoneurons during repetitive visceromotor stimulation. Bull. exp. Biol. Med. **62**, No 7, 3–8 (1966).
DUDEL, J.: Presynaptic inhibition of the excitatory nerve terminal in the neuromuscular junction of the crayfish. Pflügers Arch. ges. Physiol. **277**, 537–557 (1963).
— Potential changes in the crayfish motor nerve terminal during repetitive stimulation. Pflügers Arch. ges. Physiol. **282**, 323–337 (1965).
— KUFFLER, S. W.: Presynaptic inhibition at the crayfish neuromuscular junction. J. Physiol. (Lond.) **155**, 543–562 (1961).
DUN, F. T.: The dorsal root potential in the frog. J. Physiol. (Lond.) **95**, 41–43 (1939).
— The latency and conduction of potentials in the spinal cord of the frog. J. Physiol. (Lond.) **100**, 283–298 (1941).
— Restoration of dorsal root potential by strychnine after abolition by partial sectioning of spinal cord. Proc. exp. Biol. Med. **49**, 479–480 (1942).
— FENG, T. P.: A note on the two components of the dorsal root potential. J. Neurophysiol. **7**, 327–329 (1944).
ECCLES, J. C.: The spinal cord and reflex action. Ann. Rev. Physiol. **1**, 363–384 (1939).
— Acetylcholine and synaptic transmission in the spinal cord. J. Neurophysiol. **10**, 197–204 (1947).
— The neurophysiological basis of mind. The principles of neurophysiology, 314 p. Oxford: Clarendon Press 1953.

Eccles, J. C.: The Physiology of Nerve Cells. Baltimore: John Hopkins Press 1957.
— The nature of central inhibition. Proc. roy. Soc. B 153, 445–476 (1961a).
— The mechanism of synaptic transmission. Ergebn. Physiol. 51, 299–430 (1961b).
— Postsynaptic and presynaptic inhibitory actions in the spinal cord. In: Progress in brain research vol. 1, Brain mechanisms, ed. G. Moruzzi, A. Fessard and H. H. Jaspers, p. 1–22. Amsterdam: Elsevier Publ. Comp. 1963.
— The physiology of synapses, p. 1–316. Berlin-New York-Heidelberg: 1964 (1964a).
— Presynaptic inhibition in the spinal cord. In: Progress in brain research, vol. 12, Physiology of spinal neurons, ed. J. C. Eccles and J. P. Schadé, p. 65–91. Amsterdam: Elsevier Publ. Comp. 1964, (1964b).
— Pharmacology of central inhibitory synapses. Brit. med. Bull. 21, 19–25 (1965).
— Eccles, R. M., Magni, F.: Presynaptic inhibition in the spinal cord. J. Physiol. (Lond.) 154, 28P (1960).
— — — Central inhibitory action attributable to presynaptic depolarization produced by muscle afferent volleys. J. Physiol. (Lond.) 159, 147–166 (1961).
— Kostyuk, P. G., Schmidt, R. F.: Central pathways responsible for depolarization of primary afferent fibres. J. Physiol. (Lond.) 161, 237–257 (1962a).
— — — Presynaptic inhibition of the central actions of flexor reflex afferents. J. Physiol. (Lond.) 161, 258–281 (1962b).
— — — The effect of electric polarization of the spinal cord on central afferent fibres and on their excitatory synaptic action. J. Physiol. (Lond.) 162, 138–150 (1962c).
— Kozak, W., Magni, F.: Dorsal root reflexes of muscle group I afferent fibres. J. Physiol. (Lond.) 159, 128–146 (1961).
— Krnjević, K.: Potential changes recorded inside primary afferent fibres within the spinal cord. J. Physiol. (Lond.) 149, 250–273 (1959a).
— — Presynaptic changes associated with post-tetanic potentiation in the spinal cord. J. Physiol. (Lond.) 149, 274–287 (1959b).
— Magni, F., Willis, W. D.: Depolarization of central terminals of group I afferent fibres from muscle. J. Physiol. (Lond.) 160, 62–93 (1962).
— Malcolm, J. L.: Dorsal root potentials of the spinal cord. J. Neurophysiol. 9, 139–160 (1946).
— Rall, W.: Effects induced in a monosynaptic reflex path by its activation. J. Neurophysiol. 14, 353–376 (1951).
— Schmidt, R. F., Willis, W. D.: Presynaptic inhibition of the spinal monosynaptic reflex pathway. J. Physiol. (Lond.) 161, 282–297 (1962).
— — — Depolarization of central terminals of group Ib afferent fibers from muscle. J. Neurophysiol. 26, 1–27 (1963a).
— — — The location and the mode of action of the presynaptic inhibitory pathways on to group I afferent fibers from muscle. J. Neurophysiol. 26, 506–622 (1963b).
— — — The mode of operation of the synaptic mechanism producing presynaptic inhibition. J. Neurophysiol. 26, 523–538 (1963c).
— — — Inhibition of discharge into the dorsal and ventral spinocerebellar tracts. J. Neurophysiol. 26, 635–645 (1963d).
— — — Depolarization of the central terminals of cutaneous afferent fibers. J. Neurophysiol. 26, 646–661 (1963e).
— — — Pharmacological studies on presynaptic inhibition. J. Physiol. (Lond.) 168, 500–530 (1963f).
— Sherrington, C. S.: Studies on the flexor reflex. VI. Inhibition. Proc. roy. Soc. B 109, 91–113 (1931).
Eccles, R. M., Holmqvist, B., Voorhoeve, P. E.: Presynaptic inhibition from contralateral cutaneous afferent fibres. Acta physiol. scand. 62, 464–473 (1964a).
— — — Presynaptic depolarization of cutaneous afferents by volleys in contralateral muscle afferents. Acta physiol. scand. 62, 474–484 (1964b).

ECCLES, R. M., LUNDBERG, A.: Synaptic actions in motoneurones by afferents which may evoke the flexion reflex. Arch. ital. Biol. **97**, 199–221 (1959).

— WILLIS, W. D.: Presynaptic inhibition of the monosynaptic reflex pathway in kittens. J. Physiol. (Lond.) **165**, 403–420 (1962).

EIDE, E., JURNA, I., LUNDBERG, A.: Conductance measurements from motoneurons during presynaptic inhibition. In: Structure and function of inhibitory neuronal mechanisms, ed. C. von Euler, S. Skoglund, U. Söderberg, p. 215–219. Oxford-New York: Pergamon Press 1968.

EISEMANN, G., RUDIN, D. O.: The compound origin of potential in a stimulated dorsal root. J. gen. Physiol. **37**, 781–793 (1954).

ELLIOTT, K. A. C., FLOREY, E.: Factor I — Inhibitory factor from brain. Assay. Conditions in brain. Simulating and antagonizing substances. J. Neurochem. **1**, 181–191 (1956).

ENGBERG, I., LUNDBERG, A., RYALL, R. W.: Reticulospinal inhibition of transmission in reflex pathways. J. Physiol. (Lond.) **194**, 201–223 (1968).

FADIGA, E., BROOKHART, J. M.: Monosynaptic activation of different portions of the motor neuron membrane. Amer. J. Physiol. **198**, 693–703 (1960).

FATT, P.: Biophysics of junctional transmission. Physiol. Rev. **34**, 674–710 (1954).

— KATZ, B.: An analysis of the end-plate potential recorded with an intracellular electrode. J. Physiol. (Lond.) **115**, 320–370 (1951).

FEDINA, L., GORDON, G., LUNDBERG, A.: The source and mechanisms of inhibition in the lateral cervical nucleus of the cat. Brain Res. **11**, 694–696 (1968).

FELIX, D., WIESENDANGER, M.: Cortically induced inhibition in the dorsal column nuclei of monkeys. Pflügers Arch. **320**, 285–288 (1970).

FETZ, E. E.: Pyramidal tract effects on interneurons in the cat lumbar dorsal horn. J. Neurophysiol. **31**, 69–80 (1968).

FLOREY, E.: Comparative physiology: Transmitter substances. Ann. Rev. Physiol. **23**, 501–528 (1961).

— Amino acids as transmitter substances. In: Major problems in neuroendocrinology, ed. BAJUSZ, E., and JASMIN, G., p. 17–41. Basel: S. Karger 1964.

FORBES, A., QUERIDO, A., WITHAKER, L. R., HURXTHAL, L. M.: Electrical studies in mammalian reflexes. V. The flexion reflex in response to two stimuli as recorded from the motor nerve. Amer. J. Physiol. **85**, 432–457 (1928).

FRANK, K.: Basic mechanisms of synaptic transmission in the central nervous system. I.R.E. Trans. Med. Electronics. ME-6, 85–88 (1959).

— FUORTES, M. G. F.: Presynaptic and postsynaptic inhibition of monosynaptic reflexes. Fed. Proc. **16**, 39–40 (1957).

FRANKSTEIN, S. I., SERGEEVA, Z. N.: Presynaptic inhibition and the inhibitory Hering-Breuer reflex. Exp. Neurol. **19**, 232–235 (1967).

FRANZ, D. N., IGGO, A.: Dorsal root potentials and ventral root reflexes evoked by non-myelinated fibers. Science **162**, 1140–1142 (1968).

FUORTES, M. G. F.: Potential changes of the spinal cord following different types of afferent excitation. J. Physiol. (Lond.) **113**, 372–386 (1951).

FURSHPAN, E. J.: Neuromuscular transmission in invertebrates. Handbook of physiology, ed. J. Field, section 1: Neurophysiology, vol. 1, p. 239–254. Washington, D.C.: American Physiological Society 1959.

GALINDO, A.: GABA-picrotoxin interaction in the mammalian central nervous system. Brain Res. **14**, 763–767 (1969).

GASSER, H. S.: The control of excitation in the nervous system. Harvey Lect. **32**, 169–193 (1937).

— GRAHAM, H. T.: Potentials produced in the spinal cord by stimulation of the dorsal roots. Amer. J. Physiol. **103**, 303–320 (1933).

GERARD, R. W., FORBES, A.: "Fatigue" of the flexion reflex. Amer. J. Physiol. **86**, 186–205 (1928).

Gillies, J. D., Lange, J. W., Neilson, P. D., Tassinari, C. A.: Presynaptic inhibition of the monosynaptic reflex by vibration. J. Physiol. (Lond.) 205, 329–339 (1969).

Gobel, S., Dubner, R.: Axo-axonic synapses in the main sensory trigeminal nucleus. Experientia (Basel) 24, 1250–1251 (1968).

— — Fine structural studies of the main sensory trigeminal nucleus in the cat and rat. J. comp. Neurol. 137, 459–493 (1969).

Godfraind, J. M., Krnjević, K., Pumain R.: Doubtful value of bicuculline as a specific antagonist of GABA. Nature (Lond.) 228, 675–676 (1970).

Göpfert, H. F.: Slow potentials in the dorsal parts of the isolated spinal cord and their relation to dorsal root potentials. J. Physiol. (Lond.) 133, 433–445 (1956).

Granit, R.: The case for presynaptic inhibition by synapses on the terminals of moto-neurons. In: Structure and function of inhibitory neuronal mechanisms, ed. C. von Euler, S. Skoglund, U. Söderberg, Oxford-New York: Pergamon Press 1968.

— Kellerth, J.-O., Williams, T. D.: "Adjacent" and "remote" post-synaptic inhibition in motoneurones stimulated by muscle stretch. J. Physiol. (Lond.) 174, 453–472 (1964).

Gray, E. G.: A morphological basis for pre-synaptic inhibition? Nature (Lond.) 193, 82–83 (1962).

— Electron microscopy of presynaptic organelles of the spinal cord. J. Anat. (Lond.) 97, 101–106 (1963).

Green, D. G., Kellerth, J. O.: Postsynaptic versus presynaptic inhibition in antagonistic stretch reflexes. Science 152, 1097–1099 (1966).

Grinnell, A. D.: A study of the interaction between motoneurones in the frog spinal cord. J. Physiol. (Lond.) 182, 612–648 (1966).

Groves, P. M., Glanzman, D. L., Patterson, M. M., Thompson, R. F.: Excitability of cutaneous afferent terminals during habituation and sensitization in acute spinal cat. Brain Res. 18, 388–391 (1970).

Grundfest, H., Magnes, J.: Excitability changes in dorsal roots produced by electro-tonic effects from adjacent afferent activity. Amer. J. Physiol. 164, 502–508 (1951).

— Reuben, J. P.: Neuromuscular synaptic activity in lobster. In: Nervous inhibition, ed. E. Florey, p. 92–104. Oxford: Pergamon Press 1961.

— — Rickles, N. H.: The electrophysiology and pharmacology of lobster neuromuscular synapse. J. gen. Physiol. 42, 1301–1324 (1959).

Guzmán-Flores, C., Buendia, N., Anderson, C., Lindsley, D. B.: Cortical and reticular influences upon evoked responses in dorsal column nuclei. Exp. Neurol. 5, 37–346 (1962).

Habgood, J. S.: Antidromic impulses in the dorsal root. J. Physiol. (Lond.) 121, 264–274 (1953).

Hagiwara, S., Tasaki, I.: A study of the mechanism of impulse transmission across the giant synapse of the squid. J. Physiol. (Lond.) 143, 114–137 (1958).

Hammer, B., Tarnecki, R., Vycklický, L., Wiesendanger, M.: Corticofugal control of presynaptic inhibition in the spinal trigeminal complex of the cat. Brain Res. 2, 216–218 (1966).

Hámori, J.: Presynaptic-to-presynaptic axon contacts under experimental conditions giving rise to rearrangement of synaptic structures. In: Structure and function of inhibitory neuronal mechanisms, p. 71–80, ed.: C. v. Euler, S. Skoglund, U. Söderberg. Oxford-New York: Pergamon Press 1968.

Hancock, M. B., Willis, W. D., Harrison, F.: Viscerosomatic interactions in lumbar spinal cord of the cat. J. Neurophysiol. 33, 46 58 (1970).

Hepp-Reymond, M.-C., Wiesendanger, M.: Pyramidal influence on the spinal trigeminal nucleus of the cat. Arch. ital. Biol. 107, 54–66 (1969).

Hernández-Peón, R., Hagbarth, K. E.: Interaction between afferent and cortically induced reticular responses. J. Neurophysiol. 18, 44–55 (1955).

HOLEMANS, K. C., MEIJ, H. S.: An analysis of some inhibitory mechanisms in the spinal cord of the frog (Xenopus laevis). Pflügers Arch. **303**, 287–310 (1968a).

— — Disinhibition processes in the cord of the spinal frog. Pflügers Arch. **303**, 311–323 (1968b).

HONGO, T., JANKOWSKA, E.: Effects from the sensorimotor cortex on the spinal cord in cats with transected pyramids. Exp. Brain Res. **3**, 117–134 (1967).

— OKADA, Y.: Cortically evoked pre- and postsynaptic inhibition of impulse transmission on the dorsal spinocerebellar tract. Exp. Brain Res. **3**, 163–177 (1967).

HOUK, J., HENNEMAN, E.: Feedback control of skeletal muscles. Brain Res. **5**, 433–451 (1967).

HOWLAND, B., LETTVIN, J. Y., McCULLOCH, W. S., PITTS, W., WALL, P. D.: Reflex inhibition by dorsal root interaction. J. Neurophysiol. **18**, 1–17 (1955).

HUBBARD, J. I., SCHMIDT, R. F.: Repetitive activation of motor nerve endings. Nature (Lond.) **196**, 378–379 (1962).

— — An electrophysiological investigation of mammalian motor nerve terminals. J. Physiol. (Lond.) **166**, 145–165 (1963).

— WILLIS, W. D.: The effects of depolarization of motor nerve terminals upon the release of transmitter by nerve impulses. J. Physiol. (Lond.) **194**, 381–405 (1968).

HUDSON, R. D., WOLPERT, M. K.: Central muscle relaxant effects of diazepam. Neuropharmacology **9**, 481–488 (1970).

HUGHES, J., GASSER, H. S.: The response of the spinal cord to two afferent volleys. Amer. J. Physiol. **108**, 307–321 (1934).

IWAMA, K., KAWAMOTO, T., SAKAKURA, H., KASAMATSU, T.: Responsiveness of cat lateral geniculate at pre- and postsynaptic levels during natural sleep. Physiol. Behav. **1**, 45–53 (1966).

— SAKAKURA, H., KASAMATSU, T.: Presynaptic inhibition in the lateral geniculate body induced by stimulation of the cerebral cortex. Jap. J. Physiol. **15**, 310–322 (1965).

JABBUR, S. J., BANNA, N. R.: Presynaptic inhibition of cuneate transmission by widespread cutaneous inputs. Brain Res. **10**, 273–276 (1968).

— — Widespread cutaneous inhibition in dorsal column nuclei. J. Neurophysiol. **33**, 616–624 (1970).

— TOWE, A. L.: Cortical excitation of neurons in dorsal column nuclei of cat, including an analysis of pathways. J. Neurophysiol. **24**, 499–509 (1961).

JÄNIG, W., SCHMIDT, R. F., ZIMMERMANN, M.: Presynaptic depolarization during activation of tonic mechanoreceptors. Brain Res. **5**, 514–516 (1967).

— — — Single unit responses and the total afferent outflow from the cat's foot pad upon mechanical stimulation. Exp. Brain Res. **6**, 100–115 (1968a).

— — — Two specific feedback pathways to the central afferent terminals of phasic and tonic mechanoreceptors. Exp. Brain Res. **6**, 116–129 (1968b).

— ZIMMERMANN, M.: Presynaptic depolarization of myelinated afferent fibres evoked by stimulation of cutaneous C fibres. J. Physiol. (Lond.) **214** (in press) (1971).

JANKOWSKA, E., JUKES, M. G. M., LUND, S.: On the presynaptic inhibition of transmission to the dorsal spinocerebellar tract. J. Physiol. (Lond.) **177**, 19–21P (1964).

— — — The pattern of presynaptic inhibition of transmission to the dorsal spinocerebellar tract. J. Physiol. (Lond.) **178**, 17–18P (1965).

— LUND, S., LUNDBERG, A.: The effect of DOPA on the spinal cord. 4. Depolarization evoked in the central terminals of contralateral Ia. Afferent terminals by volleys in the flexor reflex afferents. Acta physiol. scand. **68**, 337–341 (1966).

KAHN, N., MAGNI, F., PILLAI, R. V.: Depolarization of optic fibre endings in the lateral geniculate body. Arch. ital. Biol. **105**, 573–582 (1967).

KATZ, B.: The transmission of impulses from nerve to muscle, and the subcellular unit of synaptic action. Proc. roy. Soc. B **155**, 455–477 (1962).

— MILEDI, R.: Input-output relation of a single synapse. Nature (Lond.) **212**, 1242–1245 (1966).

Katz, B., Miledi, R.: Tetrodotoxin and neuromuscular transmission. Proc. roy. Soc. B 167, 8–22 (1967a).

— — The release of acetylcholine from nerve endings by graded electric pulses. Proc. roy. Soc. B 167, 23–28 (1967b).

— — A study of synaptic transmission in the absence of nerve impulses. J. Physiol. (Lond.) 192, 407–436 (1967c).

Kawamura, H., Marchiafava, D. L.: Modulation of transmission of optic nerve impulse in the alert cat: evidence of presynaptic inhibition of primary optic afferents during ocular movements. Brain Res. 1, 213–215 (1966).

Kellerth, J.-O.: A strychnine-resistant postsynaptic inhibition in the spinal cord. Acta physiol. scand. 63, 469–471 (1965).

— Aspects on the relative significance of pre- and postsynaptic inhibition in the spinal cord. In: Structure and functions of inhibitory neuronal mechanisms, ed. C. von Euler, S. Skoglund. U. Söderberg, Oxford-New York: Pergamon Press 1968.

— Szumski, A. J.: Two types of stretch-activated post-synaptic inhibitions in spinal motoneurons as differentiated by strychnine. Acta physiol. scand. 66, 133–145 (1966a).

— — Effects of picrotoxin on stretch-activated postsynaptic inhibitions in spinal motoneurons. Acta physiol. scand. 66, 146–156 (1966b).

Kerr, F. W. L.: The ultrastructure of the spinal tract of the trigeminal nerve and the substantia gelatinosa. Exp. Neurol. 16, 359–376 (1966).

— The organization of primary afferents in the subnucleus caudalis of the trigeminal: A ligh and electron microscopic study of degeneration. Brain Res. 23, 147–165 (1970).

Khattab, F. J.: A complex synaptic apparatus in spinal cord of cats. Experientia (Basel) 24, 690–691 (1968).

Kiraly, J. K., Phillis, J. W.: Action of some drugs on the dorsal root potentials of the isolated toad spinal cord. Brit. J. Pharmacol. 17, 224–231 (1961).

Kloot,, W. G. van der: Picrotoxin and the inhibitory system of crayfish muscle. In: Inhibition in the nervous system and gamma-aminobutyric acid, ed. E. Roberts. Oxford-New York: Pergamon Press 1960.

Koketsu, K.: Intracellular slow potential of dorsal root fibers. Amer. J. Physiol. 184, 338–344 (1956a).

— Intracellular potential changes of primary afferent nerve fibers in spinal cords of cats. J. Neurophysiol. 19, 375–392 (1956b).

— Karczmar, G., Kitamura, R.: Acetylcholine depolarization of the dorsal root nerve terminals in the amphibian spinal cord. Int. J. Neuropharmacol. 8, 329–336 (1969).

Kostyuk, P. G.: Site of origin of electronic potentials in spinal roots during muscle nerve stimulation. Sechenow physiol. J. U.S.S.R. 42, 800–811 (1956).

— Presynaptic and postsynaptic changes produced in spinal neurons by an afferent volley from visceral afferents. In: Structure and function of inhibitory neuronal mechanisms, ed. C. von Euler, S. Skoglund, U. Söderberg, p. 239–248. Oxford-New York: Pergamon Press 1968.

— Preobrazhensky, N. N.: Supraspinal control of synaptic processes during viscero-motor reflexes. Bull. exp. Biol. Med. 62, (No 8), 3–7 (1966).

Levitt, M., Carreras, M., Liu, C. N., Chambers, W. W.: Pyramidal and extrapyramidal modulation of somatosensory activity in gracile and cuneate nuclei. Arch. ital. Biol. 102, 197–229 (1964).

Liley, A. W.: The effects of presynaptic polarization on the spontaneous activity at the mammalian neuromuscular junction. J. Physiol. (Lond.) 134, 427–443 (1956).

Llinás, R.: Mechanisms of supraspinal actions upon spinal cord activities. Pharmacological studies on reticular inhibition of alpha extensor motoneurons. J. Neurophysiol. 27, 1127–1137 (1964).

— A possible mechanism for presynaptic inhibition. In: Structure and function of inhibitory neuronal mechanisms, ed. C. von Euler, S. Skoglund, U. Söderberg, p. 249–250. Oxford-New York: Pergamon Press 1968.

LLOYD, D. P. C.: A direct central inhibitory action of dromically conducted impulses. J. Neurophysiol. **4**, 184–190 (1941).
— Facilitation and inhibition of spinal motoneurones. J. Neurophysiol. **9**, 421–438 (1946).
— Post-tetanic potentiation of response in monosynaptic reflex pathways of the spinal cord. J. gen. Physiol. **33**, 147–170 (1949).
— Electrotonus in dorsal root nerves. Cold Spr. Harb. Symp. quant. Biol. **17**, 203–218 (1952).
— McINTYRE, A. K.: On the origin of dorsal root potentials. J. gen. Physiol. **32**, 409–443 (1949).
LUND, S., LUNDBERG, A., VYKLICKÝ, L.: Inhibitory action from the flexor reflex afferents on transmission to Ia afferents. Acta physiol. scand. **64**, 345–355 (1965).
LUNDBERG, A.: Integration in the reflex pathway. Nobel Sympos. I: Muscular afferents and motor control (R. GRANIT, ed.), p. 275–305. Stockholm: Alquist & Wiksell. New York-London-Sydney: J. Wiley & Sons 1966.
— The supraspinal control of transmission in spinal reflex pathways. Electroenceph. clin. Neurophysiol., Suppl. **25**, 35–46 (1967).
— VYKLICKÝ, L.: Inhibitory interaction between spinal reflexes to primary afferents. Experientia (Basel) **19**, 247 (1963).
— — Inhibition of transmission to primary afferents by electrical stimulation of the brain stem. Arch. ital. Biol. **104**, 86–97 (1966).
MAJOROSSY, K., RETHÉLYI, M., SZENTÁGOTHAI, J.: The large glomerular synapse of the pulvinar. J. Hirnforsch. **7**, 415–432 (1965).
MALCOLM, J. L.: Some observations on dorsal root potentials. In: The spinal cord. Ciba Foundation Symposium, ed. G. E. W. Wolstenholme, p. 84–91. London: J. & A. Churchill, Ltd. 1953.
MALLART, A.: Heterosegmental and heterosensory presynaptic inhibition. Nature (Lond.) **206**, 719–720 (1965).
MARCHIAFAVA, P. L., POMPEIANO, O.: Excitability changes of the intrageniculate optic tract fibres produced by electrical stimulation of the vestibular system. Pflügers Arch. ges. Physiol. **290**, 275–278 (1966a).
— — Enhanced excitability of intra-geniculate optic tract endings produced by vestibular volleys. Arch. ital. Biol. **104**, 459–479 (1966b).
MATTHEWS, B. H. C.: Impulses leaving the spinal cord by the dorsal roots. J. Physiol. (Lond.) **81**, 29–31 P (1934).
McCOUCH, G. P., AUSTIN, G. M.: Postsynaptic source of dorsal root reflex. J. Neurophysiol. **21**, 217–223 (1958).
MELZACK, R., WALL, P. D.: Pain mechanisms: a new theory. Science **150**, 971–979 (1965).
MENDELL, L. M.: Positive dorsal root potentials produced by stimulation of small diameter muscle afferents. Brain Res. **18**, 375–379 (1970).
— WALL, P. D.: Presynaptic hyperpolarization: a role for fine afferent fibres. J. Physiol. (Lond.) **172**, 274–294 (1964).
MIYAHARA, J. T., ESPLIN, D. W., ZABLOCKA, B.: Differential effects of depressant drugs on presynaptic inhibition. J. Pharmacol. exp. Ther. **154**, 118–127 (1966).
MORRISON, A. R., POMPEIANO, O.: Central depolarization of group Ia afferent fibers during desynchronized sleep. Arch. ital. Biol. **103**, 517–537 (1965).
— — Depolarization of central terminals of group Ia muscle afferent fibres during desynchronized sleep. Nature (Lond.) **210**, 201–202 (1966).
NGAI, S. H., TSENG, D. T. C., WANG, S. C.: Effect of diazepam and other central nervous system depressants on spinal reflexes in cats: A study of site of action. J. Pharmacol. exp. Ther. **153**, 344–351 (1966).
NISHI, S., KOKETSU, K.: Electrical properties and activities of single sympathetic neurons in frogs. J. cell. comp. Physiol. **55**, 15–30 (1960).
NORTON, A. C.: The dorsal column system of the spinal cord (An updated review). Los Angeles: Published by: UCLA Brain Information Service 1969/70.

Pappas, G. D., Cohen, E. B., Purpura, D. P.: Electron microscope study of synaptic and other neuronal interrelation in the feline thalamus. 8th Internat. Congr. of Anatomists, Wiesbaden, 8–13 August 1965. Stuttgart: Georg Thieme 1965.

Pecci-Saavedra, J., Wilson, P. D., Doty, R. W.: Presynaptic inhibition in primate lateral geniculate nucleus. Nature (Lond.) **210**, 740–742 (1966).

Peters, A., Palay, S. L.: The morphology of laminae A and A_1 of the dorsal nucleus of the lateral geniculate body of the cat. J. Anat. (Lond.) **100**, 451–486 (1966).

Phillis, J. W.: Assay methods for transmitter substances of the central nervous system. Ph.D. Thesis, Australian National University, Canberra (1960).

— Tebēcis, A. K.: The effects of topically applied cholinomimetic drugs on the isolated spinal cord of the toad. Comp. Biochem. Physiol. **23**, 541–552 (1967).

— — Prostaglandins and toad spinal cord responses. Nature (Lond.) **217**, 1076–1077 (1968).

Pixner, D. B.: The effect of some drugs upon synaptic transmission in the isolated spinal cord of the frog. J. Physiol. (Lond.) **189**, 15P (1966).

Poritsky, R.: Two and three dimensional ultrastructure of boutons and glial cells on the motoneuronal surface in the cat spinal cord. J. comp. Neurol. **135**, 423–452 (1969).

Potter, D. D.: The chemistry of inhibition in crustaceans with special reference to gamma-aminobutyric acid. In: Structure and function of inhibitory neuronal mechanisms, ed. C. von Euler, S. Skoglund, U. Söderberg, p. 359–370. Oxford: Pergamon Press 1968.

Rall, W.: Electrophysiology of a dendritic neuron model. Biophys. J. **2**, 145–167 (1962).

— Theoretical significance of dendritic trees for neuronal input-output relations. In: Neural theory and modeling, ed. R. F. Reiss, p. 73–97. Stanford: University Press 1964.

Ralston, H. J.: The organization of the substantia gelatinosa Rolandi in the cat lumbosacral cord. Z. Zellforsch. **57**, 1–23 (1965).

Renshaw, B.: Influence of discharge of motoneurones upon excitation of neighbouring motoneurones. J. Neurophysiol. **4**, 167–183 (1941).

— Reflex discharge in branches of the crural nerve. J. Neurophysiol. **5**, 487–498 (1942).

Réthelyi, M.: Ultrastructural synaptology of Clarke's column. Exp. Brain Res. **11**, 159—174 (1970).

— Szentágothai, J.: The large synaptic complexes of the substantia gelatinosa. Exp. Brain Res. **7**, 258–274 (1969).

Rexed, B.: The cytoarchitectonic organization of the spinal cord in the cat. J. comp. Neurol. **96**, 415–496 (1952).

— A cytoarchitectonic atlas of the spinal cord. J. comp. Neurol. **100**, 297–379 (1954).

Richens, A.: The action of general anaesthetic agents on root responses of the frog isolated spinal cord. Brit. J. Pharmacol. **36**, 294–311 (1969).

Robbins, J., Kloot, W. G. van der: The effect of picrotoxin on peripheral inhibition in the crayfish. J. Physiol. (Lond.) **143**, 541–552 (1958).

Rosén, I.: Afferent connexions to group I activated cells in the main cuneat nucleus of the cat. J. Physiol. (Lond.) **205**, 209–236 (1969).

Rowe, M. J.: Reduction of response variability in the somatic sensory system by conditioning inputs. Brain Res. **22**, 417–420 (1970).

— Carmody, J. J.: Afferent inhibition over the response range of secondary trigeminal neurons. Brain Res. **18**, 371–374 (1970).

Rudomin, P.: Pharmacological evidence for the existence of interneurons mediating primary afferent depolarization in the solitary tract nucleus of the cat. Brain Res. **2**, 181–183 (1966).

— Primary afferent depolarization produced by vagal visceral afferents. Experientia (Basel) **23**, 117–119 (1967a).

— Presynaptic inhibition induced by vagal afferent volleys. J. Neurophysiol. **30**, 964–981 (1967b).

RUDOMIN, P.: Excitability changes of superior laryngeal, vagal and depressor afferent terminals produced by stimulation of the solitary tract nucleus. Exp. Brain Res. 6, 156–170 (1968).

— DUTTON, H.: Effects of presynaptic and postsynaptic inhibition on the variability of the monosynaptic reflex. Nature (Lond.) 216, 292–293 (1967).

— — The effects of primary afferent depolarization on excitability fluctuations of Ia terminals within the motor nucleus. Experientia (Basel) 24, 48–50 (1968).

— — Effects of conditioning afferent volleys on variability of monosynaptic responses of extensor motoneurones. J. Neurophysiol. 32, 140–157 (1969a).

— — Effects of muscle and cutaneous afferent nerve volleys on excitability fluctuations of Ia terminals. J. Neurophysiol. 32, 158–169 (1969b).

— — MUNOZ-MARTINEZ, J.: Changes in correlation between monosynaptic reflexes produced by conditioning afferent volleys. J. Neurophysiol. 32, 759–772 (1969).

— MUNOZ-MARTINEZ, J.: A tetrodotoxin-resistant primary afferent depolarization. Exp. Neurol. 25, 106–115 (1969).

SAUERLAND, E. K., MIZUNO, N.: Cortically induced presynaptic inhibition of trigeminal proprioceptive afferents. Brain Res. 13, 556–568 (1969).

SCHEIBEL, M. E., SCHEIBEL, A. B.: Terminal axonal patterns in cat spinal cord. II. The dorsal horn. Brain Res. 9, 32–58 (1968).

— — Terminal patterns in cat spinal cord. III. Primary afferent collaterals. Brain Res. 13, 417–443 (1969).

SCHMIDT, R. F.: Pharmacological studies on the primary afferent depolarization of the toad spinal cord. Pflügers Arch. ges. Physiol. 277, 325–346 (1963).

— The pharmacology of presynaptic inhibition. Progr. Brain Res. 12, 119–134 (1964).

— The effect of drugs on the reflex paths to primary afferent fibres. In: Studies in physiology, ed. D. R. CURTIS, and A. McINTYRE, p. 243–249. Berlin-Heidelberg-New York: Springer 1965a.

— Die Wirkung von Diazepam (Valium „Roche") auf synaptische Funktionen des Rückenmarks. Communicationes, VI. Internat. Congr. Electroenceph. Clin. Neurophysiol., Wien 1965, p. 627–630 (1965b).

— Discussion remark on the modification of cutaneous information by presynaptic inhibition. In: Touch, heat and pain, ed. A. V. S. DE REUCK, J. KNIGHT, p. 318–322. London: Churchill 1966.

— The functional organization of presynaptic inhibition of mechanoreceptor afferents. In: Structure and function of inhibitory neural mechanism, ed. C. VON EULER, S. SKOGLUND u. SÖDERBERG, p. 227–233. Oxford-New York: Pergamon Press 1968.

— Spinal cord afferents: Functional organisation and inhibitory control. In: The interneuron, ed. M. A. B. BRAZIER, UCLA Forum in Medical Sciences. Los Angeles, Calif. USA. p. 209–229 (1969).

— SENGES, J., ZIMMERMANN, M.: Excitability measurements at the central terminals of single mechano-receptor afferents during slow potential changes. Exp. Brain Res. 3, 220–233 (1967a).

— — — Presynaptic depolarization of cutaneous mechano-receptor afferents after mechanical skin stimulation. Exp. Brain Res. 3, 234–247 (1967b).

— TRAUTWEIN, W., ZIMMERMANN, M.: Dorsal root potentials evoked by natural stimulation of cutaneous afferents. Nature (Lond.) 212, 522–523 (1966).

— VOGEL, M. E., ZIMMERMANN, M.: Langsame Potentiale im Nucleus gracilis der Katze. Pflügers Arch. ges. Physiol. 291, R 4 (1966).

— — — Die Wirkung von Diazepam auf die präsynaptische Hemmung und andere Rückenmarksreflexe. Naunyn-Schmiedebergs Arch. Pharmak. exp. Path. 258, 69–82 (1967).

— WILLIS, W. D.: Intracellular recording from motoneurones of the cervical spinal cord of the cat. J. Neurophysiol. 26, 28–43 (1963a).

— — Depolarization of central terminals of afferent fibers in the cervical spinal cord of the cat. J. Neurophysiol. 26, 44–60 (1963b).

Scibetta, C. J., King, R. B.: Hyperpolarizing influence of trigeminal nucleus caudalis on primary afferent preterminals in trigeminal nucleus oralis. J. Neurophysiol. 32, 229–238 (1969).

Selzer, M., Spencer, W. A.: Convergence of visceral and cutaneous afferent pathways in the lumbar spinal cord. Brain Res. 14, 331–348 (1969a).

— — Interactions between visceral and cutaneous afferents in the spinal cord: Reciprocal primary afferent fiber depolarization. Brain Res. 14, 349–366 (1969b).

Shende, M. C., King, R. B.: Excitability changes of trigeminal primary afferent preterminals in brain-stem nuclear complex of squirrel monkey (Saimiri sciureus). J. Neurophysiol. 30, 949–963 (1967).

Skoglund, C. R., Uvnäs, B.: Phenomena in the dorsal root reflex. Acta physiol. scand. 6, 149–159 (1943).

Stewart, D. H., Scibetta, C. J., King, R. B.: Presynaptic inhibition in the trigeminal relay nuclei. J. Neurophysiol. 30, 135–153 (1967).

Stratten, W. P., Barnes, C. D.: Spinal effect of diazepam. Fed. Proc. 27, 571 (1968).

Suzuki, H., Kato, K.: Cortically induced presynaptic inhibition in cats geniculate body. Tohoku J. exp. Med. 86, 277–289 (1965).

Szentágothai, J.: Anatomical aspects of junctional transformation. In: Information processing in the nervous system, ed. R. W. Gerad and J. W. Duyff. Vol. 3, Proc. Internat. Union Physiol. Sciences, p. 119–136. Amsterdam: Elsevier, Excerpta Medica Foundation 1962.

— The structure of the synapse in the lateral geniculate body. Acta anat. (Basel) 55, 166–185 (1963).

— Synaptic structure and the concept of presynaptic inhibition. In: Structure and function of inhibitory neuronal mechanisms, p. 15–32, ed. C. v. Euler, S. Skoglund, U. Söderberg. Oxford-New York: Pergamon Press 1968.

— Hamori, J., Tömböl, T.: Degeneration and electron microscope analysis of the synaptic glomeruli in the lateral geniculate body. Exp. Brain Res. 2, 283–301 (1966).

Takagi, S. F.: The slow potential observed in the dorsal column-root preparation. Part I. On the origin of the slow potential in the spinal cord. Jap. J. Physiol. 2, 111–124 (1951).

— The slow potential observed in the dorsal column-root preparation. Part II. The concentration effects of drugs on the slow potential. Jap. J. Physiol. 4, 91–101 (1954).

Takeuchi, A., Takeuchi, N.: Electrical changes in pre- and postsynaptic axons of the giant synapse of Loligo. J. gen. Physiol. 45, 1181–1193 (1962).

Tang, A. H.: Dorsal root potentials in the chloralose-anesthetized cat. Exp. Neurol. 25, 393–400 (1969).

Tebēcis, A. K.: Effects of histamine on the toad spinal cord. Nature (Lond.) 225, 196–197 (1970).

— Phillis, J. W.: The effects of topically applied biogenic monoamines on the isolated spinal cord of the toad. Aust. J. exp. Biol. med. Sci. 45, 23–24 P (1967).

— — Reflex response changes of the toad spinal cord to variations in temperature and pH. Comp. Biochem. Physiol. 25, 1035–1047 (1968).

— — The use of convulsants in studying possible functions of amino acids in the toad spinal cord. Comp. Biochem. Physiol. 28, 1303–1315 (1969a).

— — The pharmacology of the isolated toad spinal cord. Experiments in Physiology and Biochemistry 2, 361–395 (1969b).

Therman, P. O.: Transmission of impulses through the Burdach nucleus. J. Neurophysiol. 4, 153–166 (1941).

Tömböl, T.: Short neurons and their synaptic relations in the specific thalamic nuclei. Brain Res. 3, 307–326 (1967).

Toennies, J. F.: Reflex discharge from the spinal cord over the dorsal roots. J. Neurophysiol. 1, 378–390 (1938).

— Conditioning of afferent impulses by reflex discharge over the dorsal roots. J. Neurophysiol. 2, 515–525 (1939).

TRACHTENBERG, M. C., POLLEN, D. A.: Neuroglia: biophysical properties and physiologic function. Science **167**, 1248–1252 (1970).

TREGEAR, R. T.: The relation of antidromic impulses in the dorsal root fibres to the dorsal root potential in the frog. J. Physiol. (Lond.) **142**, 343–359 (1958).

UMRATH, K.: Der Erregungsvorgang in den Motoneuronen von Rana esculenta. Pflügers Arch. ges. Physiol. **233**, 357–370 (1933).

VERHEY, B. A., KEULEN, L. C. M. VAN, VOORHOEVE, P. E.: An extreme form of presynaptic inhibition by I a afferents. Acta physiol. pharmacol. neerl. **14**, 1 (1966).

VOORHOEVE, P. E., VERHEY, B. A.: Pre- and postsynaptic effects on fusimotor- and alpha motoneurones of the cat upon activation of muscle spindle afferents by succinylcholine. Acta physiol. pharmacol. neerl. **12**, 12–22 (1963).

VYCKLICKÝ, L., RUDOMIN, P., ZAJAC, F. E., BURKE, R. E.: Primary afferent depolarization evoked by a painful stimulus. Science **165**, 184–186 (1969).

— TABIN, V.: Primary afferent depolarization evoked by adequate stimulation of skin receptors. Physiol. bohemoslov. **15**, 89–97 (1966).

WALBERG, F.: Axoaxonic contacts in the cuneate nucleus, probable basis for presynaptic depolarization. Exp. Neurol. **13**, 218–231 (1965).

WALL, P. D.: Excitability changes in afferent fibre terminations and their relation to slow potentials. J. Physiol. (Lond.) **142**, 1–21 (1958).

— The origin of a spinal cord slow potential. J. Physiol. (Lond.) **164**, 508–526 (1962).

— Presynaptic control of impulses at the first central synapse in the cutaneous pathway. In: Progress in brain research, vol. 12, Physiology of spinal neurons, ed. J. C. ECCLES and J. P. SCHADÉ. p. 92–118. Amsterdam: Elsevier Publ. Comp. 1964.

— JOHNSON, A. R.: Changes associated with post-tetanic potentiation of a monosynaptic reflex. J. Neurophysiol. **21**, 148–158 (1958).

— MCCULLOCH, W. S., LETTVIN, J. Y., PITTS, W. H.: Factors limiting the maximum impulse transmitting ability of an afferent system of nerve fibres. 3rd London Symp. on Information Theory, p. 329–344. London: Butterworth 1955.

WEISS, G. K., CRILL, W. E.: Carotid sinus nerve: primary afferent depolarization evoked by hypothalamic stimulation. Brain Res. **16**, 269–272 (1969).

WIESENDANGER, M.: The pyramidal tract. Recent investigations on its morphology and function. Ergebn. Physiol. **61**, 72–136 (1969).

— FELIX, D.: Pyramidal excitation of lemniscal neurons and facilitation of sensory transmission in the spinal trigeminal nucleus of the cat. Exp. Neurol. **25**, 1–17 (1969).

— HAMMER, B., TARNECKI, R.: Corticofugal control of presynaptic inhibition in the spinal trigeminal nucleus of the cat. The effect of pyramidotomy and barbiturates. Schweiz. Arch. Neurol. Neurochir. Psychiat. **100**, 255–276 (1967a).

— — — Corticale Beeinflussung der synaptischen Übertragung in Trigeminuskern der Katze. Helv. physiol. pharmacol. Acta. **25**, CR 237–239 (1967b).

ZIMMERMANN, M.: Habilitationsschrift, Med. Fakultät, Universität Heidelberg (1968a).

— Dorsal root potentials after C-fiber stimulation. Science **160**, 896–898 (1968b).

The Physiology of the Collateral Circulation in the Normal and Hypoxic Myocardium

WOLFGANG SCHAPER*

With 10 Figures

Table of Contents

I. Introduction

The collateral circulation of the heart is a special and extreme example of the physiology of adaptation: collateral vessels can expand to more than 10 times their initial diameter and the collateral flow can increase from virtually zero to more than 200 ml/min·100 g myocardium. This extreme degree of adaptation can furthermore proceed at an astonishing speed. In less than one

---

* Lector, Departments of Physiology and Pharmacology, University of Louvain, Belgium. Group leader, Department of Cardiovascular Research, Janssen Pharmaceutica, Beerse, Belgium.

week all necessary changes have been made to ensure normal myocardial tissue perfusion. This speed of adaptation is even more remarkable when considering the fact that the increase in vascular diameter can only be achieved by growth, i. e. active cellular proliferation, and not by passive stretch of pre-existent vascular structures. This growth process takes place in the adult organism. The collateral circulation of the heart is, however, very difficult to study. Due to limitations of the available methods, the collateral blood flow cannot be measured in man and only the very few collaterals or anastomoses which are larger than 1 mm I. D., can be visualized by cine-coronary angiography. The much more numerous and more important smaller collaterals are not seen because of the poor resolving power of present cine-angiographic techniques. In the living human heart the type of information is thus either rudimentary or inferential or indirect. After death the coronary vessels can be studied in greater detail; however, no indication as to the amount of collateral flow is available. Although the experimental restrictions are less when working with animals, the methodology is by no means easier. It is possible to measure the amount of collateral flow in the in situ beating heart, however all methods used so far have grave limitations and the results obtained with either one of these techniques should be interpreted with great caution. It is, on the other hand, completely inadequate to study only the collateral pressure flow relations, because these measurements do not offer solutions as to the question of the mechanisms of the adaptational changes. It becomes clear that the only way to tackle the problem is by a multi-disciplinary approach. This includes histology, histochemistry, cytochemistry, cell population kinetics, the study of DNA-synthesis, biochemistry and electron microscopy. The physiological side is represented by blood flow and blood pressure studies with electronic flow- and pressure-sensing devices and by the use of diffusible and non-diffusible radio-active tracers.

The problem of the collateral circulation became even more complicated and increased in urgency because surgeons introduced operative procedures for the "re-vascularization" of the myocardium and pharmacologists developed drugs which are claimed to improve the collateral circulation.

Although both ways of treatment look promising neither one has so far furnished unquestionable evidence of efficacy.

II. The Collateral Circulation in the Normal Heart
1. History, Definitions, Problems and Nomenclature

It has been known for a long time that collaterals and anastomoses are present in the human heart. RICHARD LOWER (1669) stated: "... the vessels which carry blood to the heart come together again and here and there communicate by anastomoses...".

The great adaptive ability of these vessels is also well known and Hunter who said that collaterals and anastomoses develop where and when they are needed is often quoted. These citations mostly infer that not much has been contributed since to the basic understanding of collateral development.

Over the years, particularly from the middle of the 19th century to the late thirties of this century, most investigators were preoccupied with the view that the development of large functional collaterals is dependent on the existence of small interconnecting vessels in the normal heart. This explains why so much effort was spent to show the presence (or absence) of such interarterial connections in the normal hearts of humans and other species of mammals. These research efforts were largely, if not exclusively, dependent on the more or less crude methodology. Consequently the opinions over the last hundred years showed a wave-like undulation: negative findings were followed by positive findings and so on. It is, however, strange to notice that the clear-cut evidence presented by Spalteholz (1907, 1924) in favour of the existence of collaterals in the normal human and canine heart was doubted in our time (1930–1955) by the group of Schlesinger, Blumgart and Zoll, although the information was obtained with a method definitely inferior to that of Spalteholz, which was developed almost 50 years earlier. The record was set straight again by the publication of Fulton's atlas (1965) of postmortem arteriograms ("The coronary arteries"). Fulton's beautifully detailed pictures leave very little doubt as to the existence of collaterals in the normal human heart.

However, the preoccupation with the presence of normal collaterals has distracted from the fact that functional collaterals can develop "de novo" in the absence of small arterial connections. This "neoformation" usually means a transformation of pre-existing capillaries or the transformation of new sprouting capillaries in granulation tissue. This important possibility will be extensively covered in a separate section of this review.

Although in most publications the words "collaterals" and "anastomoses" are used without discrimination some authors insist upon a clear definition of these terms. According to Schoenmackers (1958), anastomoses are interconnections between the right and the left coronary artery, collaterals are interconnections between the subbranches of the same coronary artery. This definition is definitely useful after adaptation and transformation of the vessels in question:

1. There is always a section in an anastomosis where the direction of flow is reversed.

2. When anastomoses become the main source of blood supply a larger amount of blood passes the ostium of the "donor" artery, a fact which does reduce the total coronary reserve of that artery. Example: the anterior descending artery is occluded and its distal end receives blood via anastomoses from the right coronary artery. The right coronary artery carries thus an

additional amount of blood which causes an additional pressure drop over the right coronary ostium and over the epicardial part of this artery. This reduces the total coronary reserve of that vessel because it has the effect of a "functional" stenosis.

3. When the distal end of an occluded artery is supplied with blood via collaterals originating from proximal side-branches of the same artery no additional flow is carried by the main artery.

Thus, it becomes clear that it is of some importance to distinguish between collaterals and anastomoses especially after these vessels have assumed a functional significance. In the normal heart when there is no net directional flow (and hence no "net" flow reversal) these terms are of little functional significance.

The Anglo-American literature (GREGG and FISHER, 1963) defines intra-cardiac and extracardiac vessels.

The intracardiac collateral vessels connect the coronary arteries with each other. They have been subdivided into those stemming from the same major artery, i.e. intracoronary, and those between the right and left coronary arteries, i.e. intercoronary collaterals.

The extracardiac vessels connect the coronary arterial tree with arteries outside the myocardium. Transepicardial connections are usually produced by surgical means or occasionally develop when the lungs or pleura adheres to the heart.

2. Presence, Frequency, Distribution and Function of Collaterals in Various Species of Mammals

The size and incidence of collateral vessels in the human heart is still, after 250 years of research in this field, a matter of dispute. It is not always possible to explain the large differences of opinion as differences in the experimental technique although it has recently been reported (GÖMÖRI, 1965) that small details in the methodology, such as temperature of the perfusate, position of the heart during perfusion, time after rigor mortis etc., may significantly influence the results. The great importance of appropriate perfusion pressure has been stressed by WIGGERS (1950), but unfortunately not all injection techniques are applicable at normal perfusion pressures (i.e. vinylite). Table 1 summarizes the different results obtained from studies in normal and diseased human hearts.

Two conclusions can be drawn from the literature review of Table 1.

1. Although the size of collateral vessels in normal human hearts varies in the different reports from 20–1000 μm (a factor of 50!) and although the incidence varies between 6% and 100% the more recent investigations report a higher incidence and larger anastomoses.

Table 1. *Size and frequency of coronary collaterals and anastomoses in the normal human heart and in ischemic heart disease*

Author and year	Size of collaterals		Incidence of collaterals % of cases		Number of cases
	normal hearts	ischemic disease	normal hearts	ischemic disease	
Schlesinger (1938)	40 µm and less	40–2000 µm	10%	61%	38
Zoll et al. (1951)			9%	89–100%	1050
Ravin and Geever (1946)	40 µm and less	40–2000 µm	22% of total population		166
Pitt (1959)	40 µm and less		6%	up to 100%	75
Baroldi (1956)	20–350 µm	1700–2000 µm	100%	100%	56
Laurie and Woods (1958)			75%	23%	150
Vastesaeger et al. (1958)	up to 1000 µm	1000 µm	80%	80%	120
Bellman and Frank (1958)	up to 200 µm				8
Prinzmetal (1947)	up to 170 µm				
Giese (1958)	100–200 µm	250 µm and more	about 50%	41%	71
Fulton (1963)	20–200 µm	200 µm and more	100%	100%	59
James (1961)	more than 300 µm	more than 1000 µm	100%	100%	106
Gömöri (1965)	40–100 µm	40–100 µm	more than 50%	more than 50%	128

2. Both incidence and size of the collateral vessels increase in cases of ischemic heart disease.

It seems noteworthy in this context that the incidence of collaterals is reported (Zoll et al., 1951) as being very much higher in cases of anemia even when the coronary arteries are not diseased. This was experimentally confirmed in dogs (Fam and McGregor, 1964) and pigs (Zoll and Norman, 1952).

Reiner et al. (1961) found that the hearts of human neonates are endowed with interarterial anastomoses in a high percentage of cases. This was already known to Spalteholz (1907, 1924), who interpreted this finding as a remnant of the fetal capillary network from which all larger coronary arteries evolve.

It has many times been stated (Blair, 1961; Burchell, 1940; Eckstein, 1954; Moore, 1929/30 and Pianetto, 1939), that the incidence number and size of collaterals is higher in dog hearts than in human hearts. James (1961), however, by employing the same experimental technique in the dog heart as in human hearts, found rather large and numerous anastomotic connections in human hearts but no injectable collaterals in dog hearts.

The functional significance of pre-existing collateral vessels in the dog heart is, however, very low as demonstrated by the high mortality following the acute ligation of the circumflex branch of the left coronary artery.

Only in 1 pig out of 44 small interarterial connections were found by BLUMGART et al. (1950). About the same incidence was reported by LUMB et al. (1962).

VASTESAEGER et al. (1957), however, reported injectable collaterals in 18 out of 50 pigs.

REINER et al. (1961) reported the absence of intercoronary anastomoses of about 40 µm or larger in the hearts of 29 neonatal pigs. These findings were completely identical with those of a previous series by the same authors in which the same technique was applied to the hearts of 18 adult pigs.

Our own experience showed that the same technique of postmortem arteriography which showed a rich network of anastomoses in normal dog hearts failed to demonstrate similar connections in the hearts of miniature pigs (SCHAPER, unpublished).

Our own studies in wild boar hearts which were obtained during a hunting party in the Belgian Ardennes were equally unsuccessful: no anastomoses were found after injection of a suspension of Barium-sulfate-formol. This method, however, carried out under actual "field conditions" in the woods on a cold day in January was not very sensitive.

In sheep VASTESAEGER and his colleagues (1957) were unable to demonstrate interarterial connections. Our own studies (SCHAPER, to be published) confirm the views of VASTESAEGER and his group. We found the same to be true in the hearts of neonate sheep.

3. The Function of Collaterals in the Normal Heart

FULTON (1965) and most of the authors cited in his recent monograph "The Coronary Arteries" seem to believe that those interconnecting vessels which are usually seen in the normal hearts of mammals and which are called "collaterals" or interarterial anastomoses just have no other purpose but to serve as blood conductors whenever a certain volume of blood from one artery has to be transported to another artery.

The forces necessary for such a transport (i.e. pressure gradients) are believed to be caused by the (non-synchronous) contraction of the heart muscle. It can be inferred from FULTON's text that these vessels always stay open and are functional all the time.

GREGG (1950) on the other hand stated: "The potential mechanisms for collateral reactions are the opening of pre-existing, but non-functional collaterals by an increased differential pressure, metabolites or nerve action".

This statement seems to imply that these interconnecting vessels are not necessarily patent in the normal heart, because special forces have to come into play for the "opening" of these vessels.

Another statement by GREGG and FISHER (1963) increases the importance of the foregoing because they wrote that the successful postmortal injection of an interconnecting vessel does not mean that this vessel was a patent blood conductor during life and under physiological conditions.

The finding of an endothelial inner layer in an injectable collateral vessel is, according to Fulton (1965), proof, that this vessel is a patent blood conductor under physiological conditions because the lumen of a collapsed vessel would most probably obliterate. The fate of the umbilical artery and of the ductus Botalli in mammalian neonates shortly after birth may serve as an example or analogy. Fulton (1965) assumes that the patency of coronary arterial anastomoses is maintained in the normal heart by differential blood flow. "The necessary interarterial gradients of pressure can be provided by muscular contractions. The circulatory conditions so created prevent the atrophy and disappearance of intercommunicating channels which would otherwise be their natural fate". The collateral vessels are thus assumed to be a kind of "safety exit" or safety valves to smoothen out gradients of pressure which are believed to develop during the cardiac contraction. The basis for this assumption lies in the experimental demonstration of intramyocardial pressure gradients. Johnson and di Palma (1939) demonstrated a gradient of pressure between the deeper and more superficial layers of the ventricular wall. This finding has recently been confirmed by Kirk and Honig (1964) and by Kreuzer and Schoeppe (1963).

A similar condition is believed to exist between the right and left ventricle. Because the intraventricular systolic pressure is much lower in the right ventricle than in the left the throttling of arterial inflow is much less in the right coronary artery of the dog than in the left coronary artery (Gregg, 1950).

This would also imply that a pressure gradient exists during systole between the left and right ventricular myocardium. This observation made Fulton (1965) believe "that a gradient of pressure during systole favours flow through channels which link the territories of the two arteries. This would explain the prominence of superficial anastomoses at these sites on a sound functional basis".

Although this seems to be a logical and acceptable statement there are a few but quite severe limitations to this hypothesis of the physiological role of collateral vessels.

It seems doubtful that intramyocardial pressure gradients have something to do with intravascular pressure differences. It is more likely that the squeezing actions of contracting muscle on the arteries extinguish rather than produce flow or pressure gradients. Furthermore the question arises as to whether it is necessary to extinguish possible pressure differences. And last but not least, it is not clear why pressure differences have to be diminished in the dog heart but not in the sheep heart, which lacks collaterals but certainly develops the same pressure differences.

The most fitting answer to this question points certainly in the direction of hereditary Anlage, i.e. presence or absence of collaterals is probably largely genetically determined.

If the function of collateral vessels in the normal heart as conductors of differential blood flow under the influence of pressure gradients is doubtful we must search for another explanation.

GREGG (1950), who was cited at the beginning of this chapter, seems to believe that collateral vessels are a kind of built-in safety mechanism in cases of arterial obstruction. Although FULTON (1965) is opposed to the view that collaterals are normally "quiescent", i.e. collapsed, and open only under the influence of physical and chemical factors, he does not deny the potential usefulness of the collateral vessels in coronary artery disease. However, this potential usefulness has, according to FULTON (1965), nothing to do with their normal function. FULTON (1965) develops his point of view as follows: "The statement that coronary artery anastomoses arise where they are needed (SCHLESINGER, 1938) should not be taken out of context and used as an explanation of their origin. Such an explanation is teleological and is scientifically unacceptable". "Moreover, coronary artery anastomoses have been demonstrated in lower animals in which coronary artery occlusion does not occur in the ordinary course of events".

The way out of this dilemma, i.e. the function of collateral vessels in the normal heart, is, however, possible with a very simple explanation: in accordance with the results of above cited authors, we can safely state that inter- and intra-arterial anastomoses are blood vessels of arteriolar dimensions which are normally patent and which happen to connect larger arteries with each other.

I feel that it is logical to call those vessels simply "arterioles" and ascribe to them the function of arterioles, namely to supply a capillary bed with arterial blood. Since arterioles are interconnected in a network-like structure it is quite natural that large dichotomously branching arteries are connected with each other over a network of arterioles. Collaterals are, by this definition, just the shortest connections between two arteries within this interconnecting network. According to the conventional definition, a principal distinction is made between these shortest possible connections and the rest of the arteriolar network. In my opinion this principal difference does not exist in the normal heart.

The existence of an arteriolar network is illustrated in Fig. 1, and Fig. 2 shows that the structure of an interconnecting vessel in such a network is not different from that of a normal arteriole. The properties which can rather safely be claimed for a collateral blood vessel in the normal heart are the following:

a collateral vessel has the dimensions of arterioles;

a collateral vessel has the structure of arterioles;

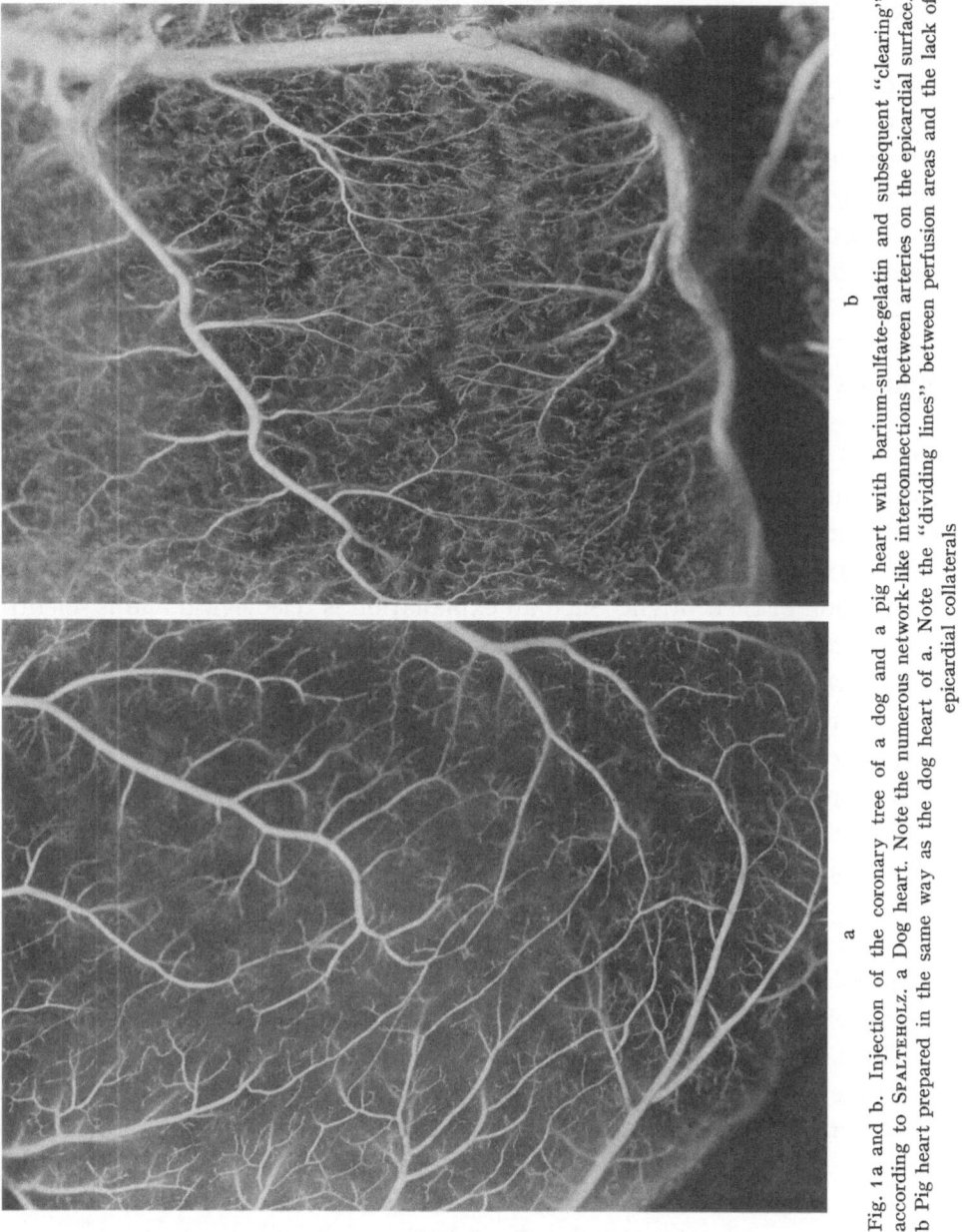

Fig. 1 a and b. Injection of the coronary tree of a dog and a pig heart with barium-sulfate-gelatin and subsequent "clearing" according to Spalteholz. a Dog heart. Note the numerous network-like interconnections between arteries on the epicardial surface. b Pig heart prepared in the same way as the dog heart of a. Note the "dividing lines" between perfusion areas and the lack of epicardial collaterals

a collateral vessel is part of an arteriolar network;

the above-mentioned structural features make it highly probable that the vessel is patent and carries blood;

a collateral vessel normally has the function of an arteriole, i.e. blood supply to a capillary bed.

The existence of an arteriolar network is undisputed and has already been demonstrated by Spalteholz in 1907, in comparison with the arterial network

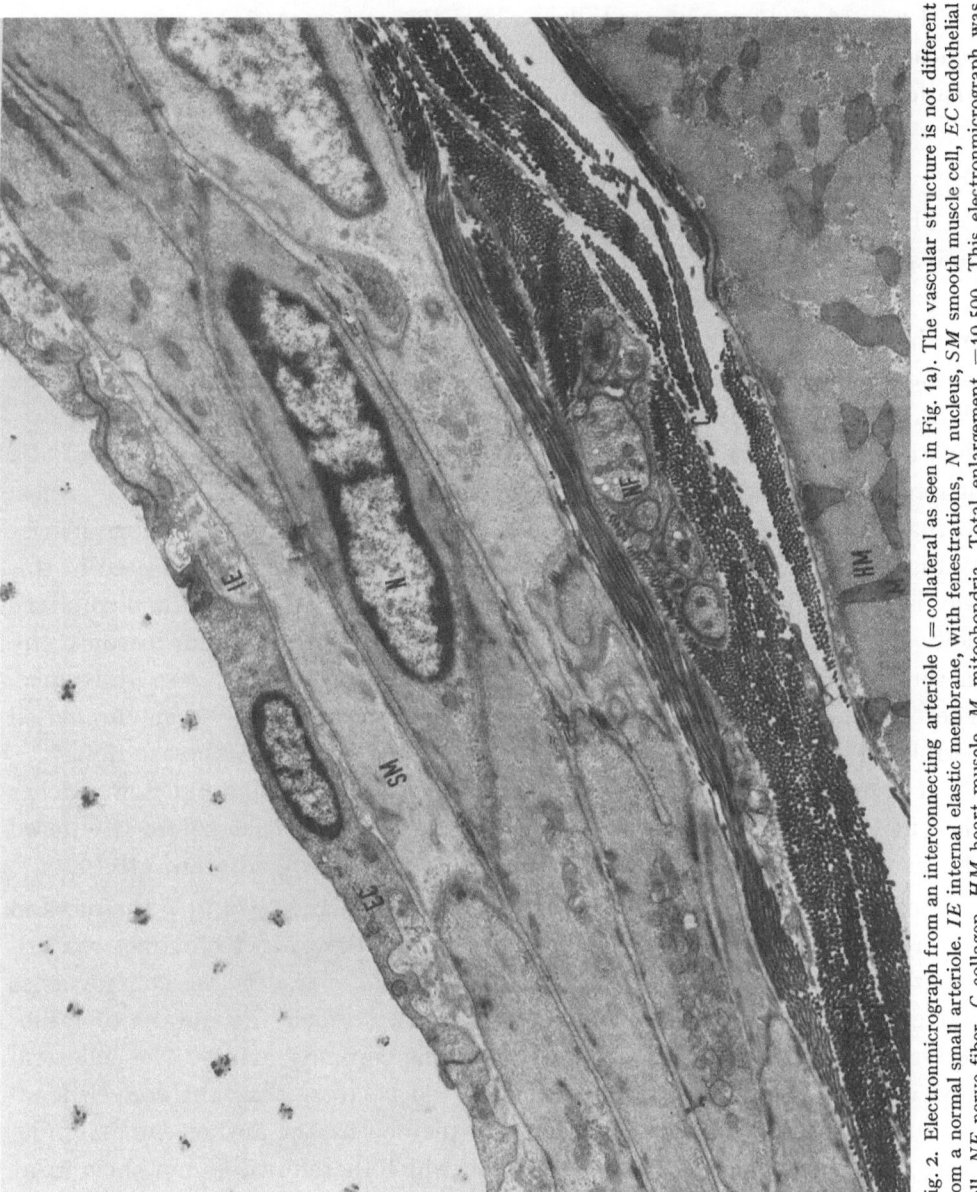

Fig. 2. Electronmicrograph from an interconnecting arteriole (= collateral as seen in Fig. 1a). The vascular structure is not different from a normal small arteriole. *IE* internal elastic membrane, with fenestrations, *N* nucleus, *SM* smooth muscle cell, *EC* endothelial cell, *NF* nerve fiber, *C* collagen, *HM* heart muscle, *M* mitochondria. Total enlargement = 10,500. This electronmicrograph was made in the author's laboratory by Dr. J. Schaper

in the skeletal muscle. I agree, however, completely with Fulton (1965) who, in discussing the problem of the "end arteries", states: "It may be conceded that the existence of small communications of 200 μm and less in diameter need not materially alter the concept of coronary arteries as end arteries from a strictly functional point of view. Denial of their existence, on the other hand, is anatomically inaccurate and doubt has been cast on the sufficiency of the evidence upon which such statements have been made".

It is of special importance to know that this statement is based on morphological results obtained in the human heart which is said to contain fewer and smaller collateral connections than the dog heart (Zoll et al., 1951).

4. Collateral Blood Flow in the Normal Heart

The methods of measurement and the actual height of the collateral blood flow are objects of controversy.

For many years the "back flow" method was widely used and accepted. Its shortcomings were known but there was nothing superior available. This method, developed by Anrep and Häusler in 1928 and widely used by Wiggers and Green (1936), Gregg (1939) and Eckstein (1941), measures the amount of blood per unit of time which leaves the peripheral cut end of a ligated coronary artery. Since the oxygen saturation of this blood is identical to that of arterial blood, it cannot have flown through nutritional capillaries and is hence true collateral blood. The total amount of blood collected in this way is about 5 ml/min. 100 g tissue which is about 6% of the normal coronary flow. This value, however, was believed to be an overestimation because the method requires that the artery bleeds freely, i.e. against zero resistance. There exists, however, a measurable pressure (10 mm Hg $\pm$ 2) in the closed peripheral artery and its reduction to zero would certainly increase flow.

It is on the other hand possible that not all of the collateral flow reaches the cut end since it could have been drained by capillaries before it entered the larger and widening arterial channels of the bleeding adjacent artery.

Because of these uncertainties as to the exact quantity of flow the method was readily abandoned, whenever a new and promising technique was advertized. Especially with the advent of radio-active tracers the controversies started. In 1961, Levy, Imperial and Zieske studied the uptake of radio-active Rubidium-86 after acute ligation of one coronary artery. The collateral flow values so obtained were surprisingly much larger than the conventional back flow values. This discrepancy was explained by the authors by assuming early capillary drainage of collateral blood which therefore does not show up at the bleeding end and by assuming a collateral circulation at the capillary level. The high flow values were, however, severely doubted when Bloor and Roberts (1965) showed that the intravascular radio-activity may markedly distort the values for Rb-86 uptake. Gregg and Fisher (1963) raised the argument that the conditions for Rb-uptake may have changed during hypoxia.

However, the argument was far from being settled because studies with inert diffusible tracers such as Krypton-85 and Xenon-133 showed also (Linder, 1966; Russell-Rees and Redding, 1967; Herd and Barger, 1964) very high values for the collateral blood flow, on the average about 20% of the normal coronary blood flow (see Table 2).

Table 2. *The coronary collateral flow acutely after ligation of an adjacent artery. K values refer to the steapness of the wash-out slopes. Flow values were not given by these authors. For further explanation see text*

Author (year)	Method	Occluded artery	Collateral flow. Mean in % of normal CBF	Collateral flow. Range in % of mean CBF	Collateral flow. Absolute values and range
Wiggers (1936)	back-flow	anterior descendens			0.5–0.7 cc/min
Gregg (1939)	back-flow	anterior descendens			<1 cc/min
		circumflex			<1 cc/min
		right			<1 cc/min
Eckstein (1941)	back-flow	anterior descendens			3.2cc/min (0.5–5.8)
Gregg (1950)	back-flow	anterior descendens			2.6 cc/min (1.9–3.4)
		circumflex			<1.0 cc/min
		right coronary artery			<1.0 cc/min
Eckstein (1954)	back-flow	circumflex			2.9 cc/min·100g (0.2–21)
Eckstein (1955)	back-flow	circumflex			3.8 cc/min (0.4–21)
Kattus (1959)	back-flow	both anterior descendens circumflex			1–3 ml/min (0–9)
Leighninger (1959)	back-flow	circumflex			4.49 cc/min (1.2–14.4)
Meesmann (1959)	thermoelectric	right coronary artery kritische Drosselung	60 %		
Leighninger (1960)	back-flow	circumflex			4.2 cc/min
Levy (1961)	back-flow	circumflex			3.53 ml/min (± 0.40)
	Rb-clearance	circumflex	13.9 %		16.3 ml/min (± 2.1)
McLean (1962)	backflow	anterior descendens			8ml/min (3–20)
Burgison (1962)	back-flow				3.3 ml/min (1.1–6.3)
Berne (1962)	back-flow	circumflex			2 ml/min

Table 2 (continued)

Author (year)	Method	Occluded artery	Collateral flow. Mean in % of normal CBF	Collateral flow. Range in % of mean CBF	Collateral flow. Absolute values and range
Lange (1963)	thermoelectric	rami ventriculares (anterior descendens)	36 %		
Betz (1964)	thermoelectric	anterior descendens	25 %		
Johansson (1964)	Kr-clearance	anterior descendens		9–20 %	5–30 ml/min
Fam (1964)	back-flow	circumflex			3.8 ml/min (1.2–11.4)
Levy (1965)	Rb-clearance	circumflex			3.47 ml/min ($\pm$ 0.55)
Pap (1965)	back-flow	anterior descendens circumflex			2.11 ml/min ($\pm$ 1.61) 1.49 ml/min ($\pm$ 1.24)
Johansson (1965)	Kr-clearance	anterior descendens		5–36 %	6–44 ml/min
Davidson (1965)	back-flow	circumflex			3.4 cc/min (1–9.6)
Bloor (1965)	back-flow H$_2^3$O-clearance	circumflex corrected for intravascular content	17 %	$\pm$ 2.5 %	12.1 ml/min ($\pm$ 2.0) 12.3 ml/min $\pm$ 2.0
		uncorrected	25 %	+ 2.4 %	19.9 ml/min $\pm$ 3.4
Rees (1966)	Xe-clearance	anterior descendens	25 %		
Schmidt (1966)	back-flow	anterior descendens			7.5 ml/min
Linder (1966)	Xe-Kr-clearance	anterior descendens main trunk	borderline 6.6 % marginal 8 %	$\pm$ 2 % + 4 %	9.1 ml/min $\pm$ 3.2 9.7 ml/min $\pm$ 4.4

Author	Method	Site	%	±	Flow
HARMAN (1966)	Rb-clearance	anterior descendens, left branch	central 9.3%, borderline 9%	+7.7%	12.9 ml/min ± 8.5, 17 ml/min
LINDER (1966)	Xe-Kr-clearance	both			9.2 ml/min, 14.8 ml/min
LINDER (1966)	Xe-Kr-clearance	anterior descendens, injection before occlusion, injection after occlusion	marginal 12.4%, central 20.8%, 20%		9.7–9.9 ml/min
REES (1967)	Xe-clearance	anterior descendens			11.9 ml/min ± 4.3, 12.6 ml/min ± 5.3, K value 0.5 (0.13–0.79)
JOHANSSON (1967)	Kr-clearance	anterior descendens	15.1%	± 5.6%	13.8 ml/min ± 5.2
LINDER (1967)	Kr-clearance	anterior descendens			12.6–22 ml/min
GRAYSON (1968)	thermoelectric	anterior descendens	10%	± 11%	
REES (1968)	Xe-clearance	circumflex	25%	7–51%	
SEEMAN (1968)	Kr-Xe-clearance	anterior descendens			11.2–11.3 ml/min
REES (1969)	Xe-clearance	anterior descendens	8%		K 1.67 before, K 0.21 after occl.
REES (1969)	Xe-clearance	anterior descendens	23%	8–32%	
HAFT (1969)	Kr-clearance	both, plug occlusion, ligature	30.2%, 27.8%		
EDLICH (1969)	Rb-clearance	anterior descendens			40 ml/min

However, the inert gas methods have some serious shortcomings which limit their use. DOUTHEIL (1966) has shown that the conventional analysis of the disappearance-slope is inaccurate because of the contribution of non-cardiac radio-activity. Although radio-Xenon as well as radio-Krypton are exhaled during the first passage through the lungs, the extracardiac radio-activity can influence the scintillation counter which is usually not very selectively collimated. It is therefore preferable to use highly selective "spot-measuring" collimating systems such as the Honey-cone collimator (HUSAK and PERINOVA, 1969), which is insensitive to all non-cardiac radio-activity. Selective collimation, however, reduces the efficiency and sensitivity of counting so that much higher radio-active doses have to be injected.

It is furthermore mandatory that the injected tracer almost instantaneously saturates the muscle. This was shown to be the case by CANNON et al. (1969) with the γ-scintillation camera when the tracer was injected slug-like into a normal coronary artery. After coronary artery occlusion, however, the situation is completely different. Since there is practically no flow in the peripheral blind end of the artery, the injected tracer just remains there as a depot. The wash-out curve is therefore nothing else than the slow disappearance of the depot plus the almost simultaneous "wash in" and "wash out" to and from the muscle but sometimes also backward leakage of the tracer through a not too tightly fitting stopcock!

Using cine-coronary arteriography after injection of Urografin[1] into the peripheral end of the ligated artery, we found depot-formation and very slow depot-dilution even in hearts with a very well developed collateral circulation (SCHAPER et al., unpublished). This means that repeated measurements of collateral flow through repeated injections into the peripheral end of a ligated artery should be regarded with utmost suspicion!

Well aware of the apparent limitations of the method and largely eliminating them by a careful choice of experimental conditions, LINDER et al. (1966) measured the collateral flow in small areas of myocardium and came to several important conclusions:

1. the collateral flow is inhomogeneously distributed. It is fairly high near the border zones of the hypoxic area and falls to practically zero in the center of the hypoxic zone.

2. The vessels located in normal myocardium but adjacent to the ischemic zone are dilated but they still retain some tone.

3. The collateral flow is influenced by the perfusion pressure, heart rate and by myocardial stretching.

Our own collateral flow studies were started with the idea in mind to show differences between different species of mammals. We firmly believed before doing the experiment that a tremendous difference must exist between the

---
1 Meglumine Diatrizoate.

amount of collateral flow in the dog and that of the sheep the latter animal lacking collaterals completely.

We used dogs, pigs and sheep in this study and tried to observe all necessary precautions: we used a "spot-measuring" critically focussed Honey-cone collimator, the chest was opened and only the eventually and intermittently hypoxic myocardial area was scanned. Furthermore, the tracer was always injected in the non-occluded artery and was "washed in" into the muscle by the blood stream. Thereafter the artery was occluded during 90 sec. The Xenon-133 disappearance curves obtained in this way showed some typical features. Wash-out during arterial clamping was visibly slowed down and was markedly accelerated after re-opening of the artery (= reactive hyperemia). Zero levels were again (and always) reached shortly after the reactive hyperemia indicating that no radio-active background was piling up. Although the experiments looked correct from the technical side we were astonished to see that after analyzing the data no systematic difference was found between the collateral flows of dogs, pigs and sheep. The lack of collaterals in sheep and pigs and their presence in dogs was confirmed by postmortem injection studies in the same hearts. This unexpected result is very difficult to explain and remains an enigma to us.

Another possibility to study collateral flow is the direct injection of Xenon into the heart muscle (LINDER, 1966). Preliminary experiments in our laboratory with this technique in the dog showed good agreement with intravascular Xenon injections when the artery was not occluded. After occlusion, however, the clearance of intramuscular Xenon was much slower than after previous saturation via the artery. The result is, of course, dependent on the injection site because the collateral flow is non-homogeneous as shown by LINDER et al. (1966). The clearance from intramuscular fluid pockets has, however, serious disadvantages: injection of small aliquots with a thin needle may cause leakage after withdrawal of the needle and measurements are seldom repeated at the same spot. The last difficulty is overcome by inserting a thin plastic catheter with side-holes into the myocardium. With this technique, however, one wonders what happens to the capillaries surrounding the implant: they may be compressed, do not carry blood, or they may be damaged and bleed into the pocket.

Another very promising new technique was developed by DOMENECH et al. (1969), RUDOLPH et al. (1967) and NEUTZE et al. (1968), who used radioactively labelled microspheres of different diameter (15–50 μm) which are injected into the left atrium and are proportionately distributed with the cardiac output. By comparing the radio-activity of the recovered fraction (organ sample) with the total amount of injected radio-activity the organ flow can be calculated. Because of the sensitivity of present day gamma well-counting only very minute amounts of microspheres need be injected.

No hemodynamic changes due to the obstruction of pre-capillary vessels have been observed because approximately only one out of thousand small arterioles is obstructed after about 4 injections. No absolute values of collateral flow have been reported so far, only area-comparisons in the form of cpm-ratios were published by BECKER et al. (1969).

III. The Vascular Adaptation to Chronic Hypoxia

Great uncertainty exists as to the definition, meaning and significance of the term "chronic hypoxia", which is as frequently used as "chronic ischemia" or chronic "coronary insufficiency". For the heart muscle, lack of oxygen can never be a chronic condition because of its very limited tolerance of anoxia. What is meant is the gradual narrowing of a major coronary artery until complete occlusion, which includes a phase of myocardial hypoxia of variable duration. Because the vascular system reacts to hypoxia with a functional and structural dilatation the result of the hypoxic phase is either salvage of cardiac muscle, because of collateral enlargement, or myocardial infarction in case of failure of the collaterals to enlarge.

From our own studies in vascular adaptation and from the studies of several other authors (GREGG, 1950; ECKSTEIN, 1955, 1957), it became obvious that the degree of hypoxia has to be delicately balanced between the danger of infarction and the necessities of the adaptational process.

It did therefore not surprise us when we found that an acute experimental myocardial infarct is not the best stimulus for collaterals to enlarge. In such an experimental model the hypoxia is extremely intense; however, the muscle is dead after one hour or so and dead muscle does not deliver the stimulus of hypoxia. It is our experience from more than one hundred acute experimental infarctions that the collateral circulation develops very slowly and reaches a degree inferior to that usually observed after slow coronary artery occlusion. To fully explore the potential of the collateral adaptation we used therefore the Ameroid technique developed by LITVAK (1957) and VINEBERG (1960). The Ameroid constrictor consists of a hygroscopic plastic material which increases its volume when taking up tissue fluid. Since the outer dimensions are kept constant by encapsulation in a stainless steel ring the constrictor, when slipped over a coronary artery, will narrow the vessel. The time until complete occlusion can be controlled to a certain degree. It can be accelerated so that the artery is occluded within 48 h (LUMB, 1963) by using a device with an eccentric bore, by closing the gap through which the artery was introduced with a slot, by placing the constrictor in a desiccator 24 h before implantation, by using a constrictor somewhat smaller than the artery at the moment of implantation. Constriction can be delayed by using different kinds of Ameroid plastic and by treating it with mineral oil or white purified vaseline.

In our experiments we used a type which occluded the circumflex coronary artery of dogs and sheep and the anterior descending artery of pigs within $2^1/_2$ weeks after the implantation.

1. Death Rate and Myocardial Infarction after Coronary Artery Occlusion

Mortality after coronary artery occlusion is very variable and numerical values obtained in one laboratory cannot be compared with those in another. The discrepancy between figures is extreme in the case of the mortality following acute ligation of a coronary artery: it was reported to vary between 0 and 100 %. In my laboratory the mortality of the anterior descendens occlusion is now 10 %. The ligation of the larger left circumflex coronary artery carries a higher mortality: in a series of 40 unselected mongrel dogs only 20 % survived the first hour after ligation. When the circumflex coronary artery was slowly constricted a developing collateral circulation increased the number of surviving dogs to about 70 %. These values, however, apply only to a group of unselected mongrels with an average body weight of about 15 kg and of unknown age. Surprisingly enough young purebred beagle dogs below the age of 3 months survive even acute ligation of the circumflex coronary artery and all adult beagles older than one year survived the slow constriction of this artery (SCHAPER, unpublished observations). Survival of the adult beagles is not necessarily related to a better collateralization because we found that the incidence of infarcts was almost as high as in the mongrel population. Quantitative assessment showed, that beagles are less prone to post-infarction arrhythmias than are mongrels (SCHAPER and XHONNEUX, unpublished).

In mongrels, on the other hand, we found a good correlation between the degree of postinfarction arrhythmia and the degree of postinfarction collateralization. The larger the pre-existent collateral network the lesser the number of postoperative ventricular premature beats.

On the other hand, a selected population of old mongrels had the same postoperative mortality from Ameroid-constriction as the non-selected total population of mongrels (SCHAPER, to be published).

The dogs which died around the time of the estimated date of complete coronary artery occlusion by Ameroid-constrictor usually died of ventricular fibrillation which we were able to record several times. Ventricular fibrillation was observed in these dogs to happen during excitement, after a meal, during a fight with other dogs, or during the walk from the kennel to the laboratory.

Death because of cardiogenic shock or because of heart failure was never observed after Ameroid-constriction in dogs; it was, however, a common observation in sheep and, to a lesser extent, in pigs.

The frequency of myocardial infarction after Ameroid-constriction varied also between species.

If we assume that animals dying from ventricular fibrillation did so because of myocardial infarction, we have an incidence of myocardial infarction of about 50% in dogs. Thirty per cent died suddenly, 20% survived with a clearly demonstrable infarct and about 50% of all dogs survived complete coronary artery occlusion without an infarct.

In dogs the infarct was always located in the area of the posterior papillary muscle. The infarct was never transmural, comprised seldom more than $1/_3$ of the thickness of the left ventricular wall and occupied often only the top of the papillary muscle.

In pigs the situation is somewhat different. Although we reported in an earlier paper (Schaper, Jageneau and Xhonneux, 1967) that about 25% of the pigs with an Ameroid-induced occlusion survived without myocardial infarct we have to correct that figure, because of new findings from our laboratory. This 25% was largely obtained by macroscopic inspection and several infarcts may have been overlooked. A microscopical review of our last series of 50 carefully analyzed pigs revealed that less than 10% survived without a myocardial infarct. The infarct in pigs is mostly transmural and the connective tissue which replaced the heart muscle shows an early tendency to calcification.

All of a series of 9 sheep with Ameroid-constriction of the circumflex artery developed large transmural infarction and died of heart failure or of cardiogenic shock.

2. The Peripheral Coronary Pressure as an Index of Collateral Development

When the left circumflex coronary artery is acutely occluded the diastolic pressure in the peripheral part of the occluded artery falls within about 10 heart beats after ligation to a value approximating 8–10% of its original value. This pressure is called peripheral coronary pressure (PCP) or back pressure (Gregg, 1950; Schaper, 1967). The PCP is pulsatile and, when sufficiently amplified, looks like the left ventricular pressure tracing. The PCP is known to rise some time after coronary artery constriction or occlusion. However, the exact time relations during and after slow constriction have never been quantitatively determined in a larger population of animals. Our own material is sufficiently large (110 dogs) to be subjected to a statistical analysis which will be presented in this review. The rise of the PCP after coronary artery occlusion is, of course, the result of collateral enlargement. When the collaterals and anastomoses enlarge, the transmission of the aortic pressure becomes better and better until, after a sufficiently long time, the PCP is practically as high as the coronary perfusion pressure.

The dependency of the PCP on the aortic pressure is demonstrated in Fig. 3. The arterial blood pressure was varied in a sinusoidal way using the

pump described by WETTERER and PIEPER (1953). The aortic pressure oscillations caused similar in-phase oscillations of the PCP. Because of this relation
between the PCP and the aortic pressure and because of the fact that coronary
pressure-flow relations are least effected by extravascular factors only at the
end of diastole, the diastolic PCP and the diastolic aortic blood pressure were
always expressed as a dimensionless ratio. This reduced the scatter caused by
the different values of the aortic pressure within a population of animals.

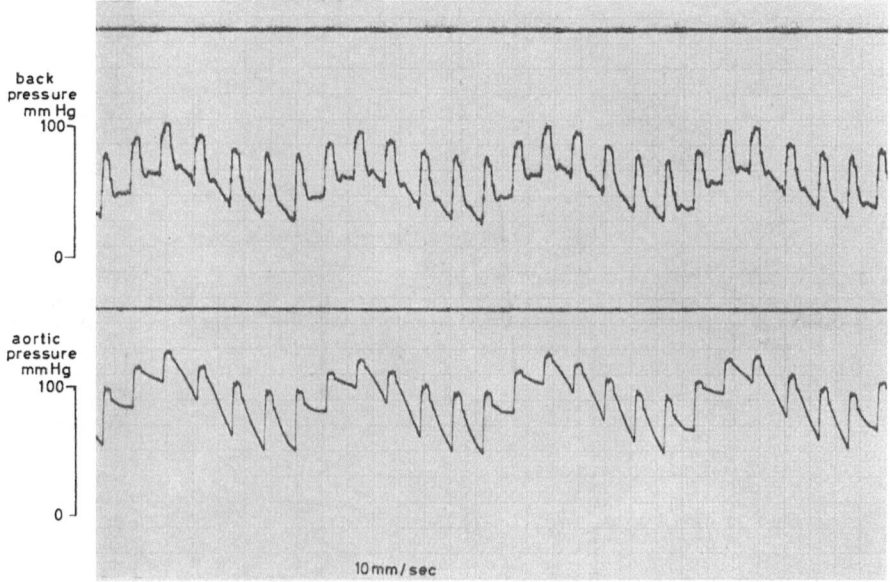

Fig. 3. Influence of forced oscillations of the arterial blood pressure (lower panel = aortic blood pressure)
on the PCP or back pressure. Note how well the PCP follows the aortic pressure oscillations which were
induced by with-drawing and re-injecting blood from and to the arterial system in a sinusoidal way

The time course of the PCP measured at various time intervals after the
implantation of an Ameroid-constrictor is illustrated in Figs. 4–7.

During the first 2 weeks after the occlusion the back pressure rises very
sharply, the steepest part of the curve being about 12 mm Hg/day. After the
4th week of implantation of the constrictor the curve begins to level off. There
is very little, but statistically significant, pressure increase between 3 months
and half a year but there is no further increase between half a year and one
year. Half a year after the implantation of the constrictor the ratio diastolic
back pressure versus diastolic aortic pressure (= 0.85) has come close to the
theoretical maximum, namely 1.0. Of the 11 6-month dogs, 6 had pressure
ratios at or above 0.90. The average value of the 6-month period was influenced by 2 unusually low values and the substitution of the average value
by the median value would have resulted in a ratio of 0.90. The conversion of
the actual back pressure values into dimension-less ratios was done because of

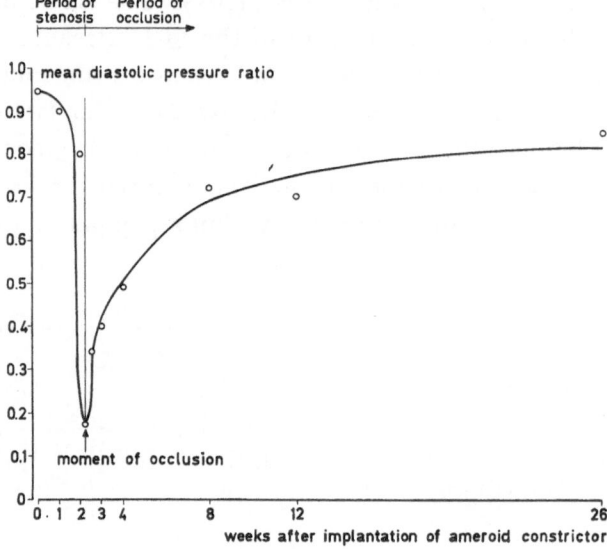

Fig. 4. Time course of the pressure measured intra-arterially and downstream from the Ameroid constrictor. Values are expressed as the ratio between late diastolic PCP and late diastolic aortic blood pressure. The ratio falls as a result of the arterial stenosis and the ratio rises again (after complete occlusion) as a result of collateral enlargement

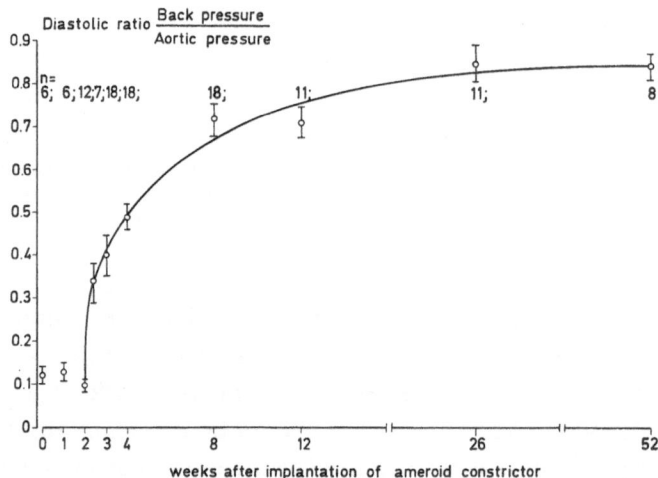

Fig. 5. Ratios of late diastolic PCP over late diastolic ABP as a function of time. Fig. 5 shows that the ratios at 1 and 2 weeks are very low (obtained after occluding the already stenosed artery) and not different from values obtained by acute ligation in normal dogs. N size of the population studied at that moment

the direct influence of the aortic pressure on the back pressure (SCHAPER, 1967, 1966; see also Fig. 3).

When the stenosed artery (1 and 2 week groups) distal from the constrictor was not occluded by a ligature the pressures were, of course, considerably higher. The fall of the poststenotic pressure as a function of time is illustrated in Fig. 4. As a result of arterial narrowing and of poststenotic compensatory dilatation of the larger and, more important, of the arteriolar resistance-vessels the pressure drops as the stenosis progresses.

The lowest pressure ratio of Fig. 4 has been calculated from a small group of dogs with occluded arteries from the groups 2, 2.5 and 3 weeks with the lowest back pressures ever observed. These might closely approach the lowest possible back pressure at the moment of occlusion. From this moment on the pressure rises again rather steeply due to the growth of collateral vessels (SCHAPER, 1967). This part of Fig. 4 is, of course, identical with the corresponding part of Fig. 5.

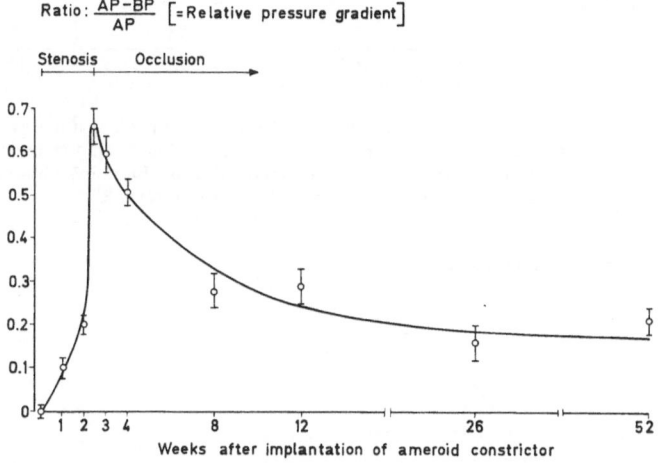

Fig. 6. Time course of the pressure gradient over the collaterals. AP late diastolic aortic pressure measured closely to the coronary ostium; BP back pressure $(= PCP)$

Fig. 6 illustrates the time course of the pressure gradient, which rises and falls as a mirror image of Fig. 4 from which it was partly derived. Fig. 6 shows that the pressure gradient over the collateral vessels rises already during the period of stenosis and falls during the period of collateral enlargement.

3. Critique of Methodology, Definitions and Assumptions

The main disadvantage of producing a slowly developing chronic coronary artery occlusion with the Ameroid technique (LITVAK et al., 1957) is the fact that the exact moment of coronary artery occlusion is not known. Even a minute amount of mismatch between the size of the artery and the size of the constrictor at the time of implantation can cause deviations of several weeks from the predicted moment of occlusion.

These deviations were reduced to a minimum by strictly observing the following rules:
all constrictors were made in our own shop by the same instrument maker and from the same batch of plastic material;

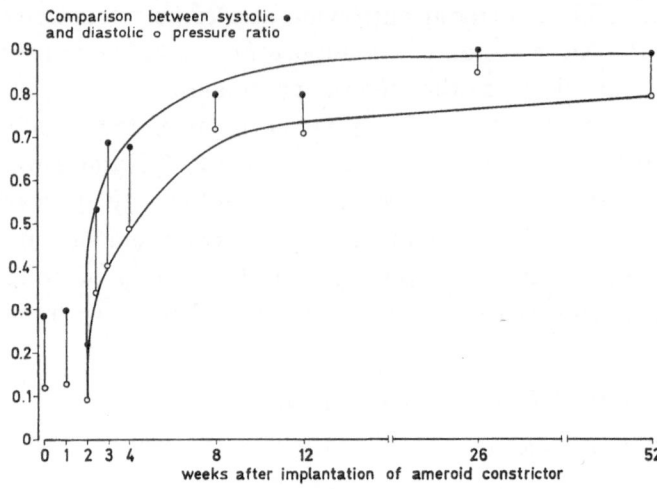

Fig. 7. Comparison between systolic and diastolic PCP/ABP ratios. The peak systolic PCP was divided by the peak systolic ABP. Note that the systolic ratios are always higher than the diastolic ratios. Note furthermore that the systolic-diastolic ratio-difference is largest during the rising phase of the curve and smallest when the curve levels off at later intervals

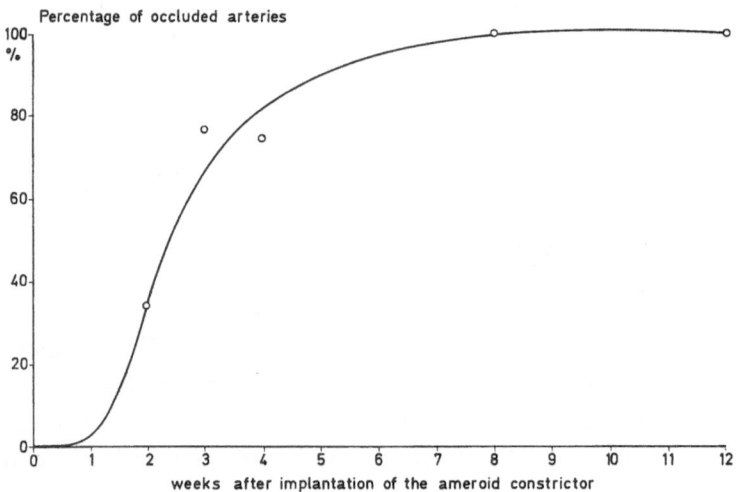

Fig. 8. Percentage of occluded arteries (by Ameroid constrictor) as a function of time. Statistical analysis (by probit analysis) showed that the average time to occlusion is $2^1/_2$ weeks

all Ameroid-constrictors were kept in very dry air in a tightly stoppered bottle in which the drying material (silica-gel) was renewed regularly;

all constrictors were implanted by the same surgeon who carefully matched the arterial size with the constrictor size intra operationem.

By observing these rules, extreme variations in the time course to occlusion could be avoided. With our own constrictors and with the above described technique more than 84% of all constrictors occluded the artery between the 2nd and the 3rd week after implantation (see Fig. 8). In an attempt to further reduce the scatter of values, all dogs with an already occluded artery 2 weeks after the implantation were called the "2.5-week group", and all dogs with a stenosed artery 3 weeks and more after the operation were rejected.

The pressure gradient is defined as follows: the pressure gradient is the pressure difference over the lumped collateral system per unit collateral length. Since the length of individual collaterals is so variable and since the unit length of the lumped collateral vessels is a rather awkward term we simply calculated the pressure difference between the aortic pressure, which was measured close to the orifice of the left coronary artery, and the pressure in the distal end of the occluded artery. It was assumed that the aortic pressure is identical with the pressure at the very beginning of a collateral vessel and that the back pressure is equal to the pressure existing at the re-entrance region of the collateral vessel. Both assumptions do not deviate more than 10% from the true pressure gradients which had been obtained by micropuncture studies in 3 dogs (unpublished work).

Fig. 6 shows that the pressure gradient rises and falls as a mirror image of Fig. 4.

This is quite understandable since:

$$\frac{AP-BP}{AP} = 1 - \frac{BP}{AP} \tag{1}$$

It is, on the other hand, quite conceivable that the back pressure mounts as the resistance of the collateral vessels decreases (see also network theory). Since the pressure gradient is only a derivative of the back pressure, only a parallelity between the enlargement of collaterals and the pressure gradient can be assumed but not a causal action of the pressure gradient as has been postulated by GREGG and FISHER (1963).

4. Simulation of the Collateral Resistance by Electrical Networks

Fig. 9 shows a schematic presentation of the left coronary artery with one of its two main branches occluded by the constrictor. It is quite suggestive that this model resembles the unbalanced Wheatstone bridge with one of the main feeding resistors showing a very high resistance.

The proposed model of Fig. 9 is based on the assumption that fluid-flow through a system of arteries under the action of pressure gradients is comparable to the flow of electric current through electrical resistances under the action of potential differences and that Ohm's law applies as well for the hydraulic model as for the electronic model.

The hemodynamics of diastolic blood flow through the coronary system may be carried out safely in terms of laminar stationary flow, when only mean diastolic flow and pressure are to be considered. Hence the flow pattern is completely determined by the distribution of pressures and by the magnitudes

of resistances. When Poiseuille's law for laminar stationary flow is applied we
will obtain a set of linear equations between flow and pressure gradients. These
equations are formally identical to Ohm's law for an electrical conductor.
Speaking in terms of electrical analogs has an advantage over the use of hemodynamical magnitudes, because of the progress that has been made in circuit
analysis and in the application of digital and analog calculators.

The preference of electrical above hemodynamical analogs is more accidental
than fundamental. At the time when Ohm made his famous discovery, electrical phenomena were currently described in terms of hydrodynamical
magnitudes.

The above mentioned assumptions of laminar flow and "dc"-conditions
are acceptable especially when the back pressures and aortic pressures were

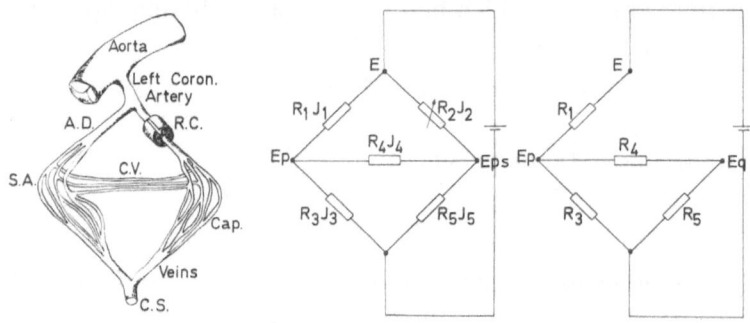

Fig. 9. Network model of the coronary circulation. Substitution of the coronary anatomy by electrical
symbols. For further explanation see text

measured in late diastole, when phase-shifts between perfusion pressure and
fluid movement are minimal (Gregg and Fisher, 1963).

The simplest possible resistive model for the coronary collateral circulation
is represented in Fig. 9. The conversion of physical into electrical magnitudes
proceeds as follows:

R_1, I_1 resistance and blood flow through the anterior descending branch of
the left coronary artery;

R_2, I_2 resistance, flow, through the circumflex branch of the left coronary
artery;

R_3 combined arteriolar and capillary resistances of the area perfused by
the anterior descendens artery;

R_4 lumped collateral resistance;

R_5 combined arteriolar and capillary resistance of the area perfused by
the circumflex artery.

The back pressure and poststenotic pressure are measured at the point
Eq and Eps respectively.

Since R_1 is assumed to be very small as compared to R_3, R_4 and R_5, the value measured at point Ep in the network is assumed to be equal to the diastolic aortic blood pressure.

The ratio of back pressure to aortic pressure at late diastole (Eq/Ep) can be expressed in terms of resistances of our network model. After solving the Kirchhoff equations we find:

$$\frac{Eq}{E} = \frac{R_3 - R_5}{(R_4 + R_5)(R_3 + R_1) + R_1 \cdot R_3} \cdot \tag{2}$$

Assuming that the ratio of arterial (R_1) to arteriolar (R_3) resistance is small with respect to unity, we may express the relation as:

$$\frac{Eq}{E} = \frac{R_5}{R_4 + R_5} = \frac{BP}{AP} \cdot \tag{3}$$

This means that the ratio of back pressure to aortic pressure can be expressed in terms of collateral (R_4) to arteriolar (R_5) resistances, i.e. the back pressure is influenced by the resistance of the collateral vessels but also by the resistance of the arteriolar resistance vessels which the collaterals supply with blood. By means of hydraulic model studies, ECKSTEIN (1954) has also shown that the back pressure is influenced by the peripheral run-off.

The equation (3) shows quite clearly that back pressure:aortic pressure ratios can approach their theoretical maximum (i.e. 1.0) only when R_4 (i.e. the collateral resistance) becomes very small. However, a dilatation of all resistance vessels (R_5) at R_4-values which had not yet approached their lowest possible values would result in a fall of the BP/AP-ratio which is indeed the case when a coronary vasodilator is injected (SCHAPER, 1967). The rate and degree of change of the collateral resistance can be estimated by comparing the results contained in Figs. 4 and 5 with calculations based on the Wheatstone bridge model.

Starting from equation (3)

$$\frac{Eq}{E} = \frac{R_5}{R_4 + R_5}$$

we call the back pressure ratio α (i.e. $BP/AP = Eq/E = \alpha$). Then we obtain

$$R_4 = R_5 \frac{1 - \alpha}{\alpha} \cdot \tag{4}$$

We have now to distinguish between 2 cases:

a) The collateral resistance R_4 has not yet decreased to such a value that a normally required blood flow over R_5 is obtained; we clearly may here assume that R_5 takes a minimal value $(R_5)_{min}$ and our relation (4) is written as

$$R_4 = (R_5)_{min} \frac{1 - \alpha}{\alpha} \cdot \tag{5}$$

b) R_4 reaches a value that restores a normal perfusion over R_5. This critical value of R_4 will be called $(R_4)_c$ and a necessary condition appears

$$(R_4)_c + (R_5)_{min} = (R_5)_0 \cdot \tag{6}$$

When R_4 decreases beyond this critical value $(R_4)_c$, the arteriolar R_5 is allowed to increase again, which is expressed by

$$R_4 + R_5 = (R_5)_0.$$

Inserting this condition into our basic relation (4), we obtain

$$R_4 = (R_5)_0\,(1-\alpha). \tag{7}$$

We now have a set of two equations to describe the radial enlargement of the collateral vessels with time, starting from experimental Eq/Ep-curves.

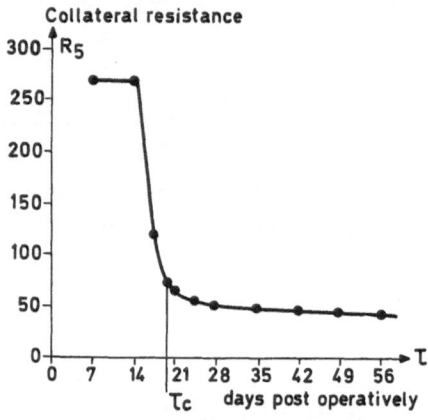

Fig. 10. Computer plot showing the fall of the collateral resistance, based on calculations carried out with the resistance model of Fig. 9

The critical reversal point on the time scale is easily determined, by equating relations (5) and (7). It is evident that the two equations must match at the reversal point. Hence we find

$$\alpha_c = \frac{(R_5)\mathrm{min}}{(R_5)_0}. \tag{8}$$

The value of α_c must be near 0.2, which has been shown experimentally (see Fig. 4).

From the Eq/Ep versus time curve, we find that reversal occurs after about 20 days, i.e. just before completion of the third week. Fig. 10 summarizes the results of calculations, which are reduced to a normal arteriolar resistance R_5 equal to 100.

Although I have tried to use more complex networks which necessitated the use of a digital computer[2] (Schaper, 1967) the information so obtained did not differ principally from that of the above described simpler network.

The most important contribution of these network studies to the understanding of the PCP is the insight that the PCP is always influenced by the

2 These studies were carried out by Paul Lewi, head of the computer centre of the Research Laboratoria, Janssen Pharmaceutica, Beerse, Belgium.

ratio of 2 resistances, the lumped collateral resistance and the resistance of the arterioles which receive their blood via collaterals. In this ratio of resistances only one (the arteriolar) can change its value acutely. The other (collateral) has a relatively fixed value which can only change by morphological alteration, i.e. growth. The possibility of the acute change of the arteriolar resistance can heavily influence the PCP which is easily demonstrated by the injection of a vasodilator drug. Arteriolar dilatation and increased collateral flow lead to an accentuated pressure fall over the lumped collateral resistance which decreases the PCP. This must not be confused with the situation which leads to a rise in PCP with increasing collateral flow which takes place during growth of the collateral vessels.

It becomes clear that the PCP, although it is a very useful indicator, cannot fully explain the hemodynamic situation without an estimation of the magnitude of the collateral blood flow.

Another shortcoming of the above described network is the assumption that the artery which supplies the collaterals with blood, i.e. the anterior descending branch of the left coronary artery in our model, offers only a negligible resistance. This is true for most practical purposes in non-diseased arteries. It is, however, doubtful whether this holds also for multiple vessel disease in the human heart. An analysis of this situation becomes much more complex and necessitates a large scale computer simulation. It can, however, already be said that induced general coronary vasodilation may decrease the collateral flow when the anterior descending "donor" artery is also stenosed.

5. The Measurement of the Collateral Flow during and after Enlargement of Collaterals

The measurement of the collateral flow in chronic coronary artery occlusion poses great technical difficulties. The main difficulty is the uncertainty of adequate myocardial saturation with the tracer gas when the tracer was injected into the occluded peripheral end of the artery. These difficulties may have prevented studies which show day-to-day changes of collateral flow during the process of slowly progressing coronary artery stenosis-occlusion.

As explained earlier in this review the injection of a bolus of radio-Xenon into the peripheral end of an occluded artery may result in the formation of an arterial depot. Such studies are relevant only when the anatomical situation, i.e. the existence of large high-basal anastomoses, favors the quick saturation of muscle with the tracer. We know from experience that this may not be the case in at least 50 % of all animals studied even in the presence of a very well developed collateral circulation.

When measuring collateral flow some time after coronary artery occlusion, one usually finds normal tissue perfusion provided that serious infarction had not occurred. Normality of tissue perfusion, however, does not give information about the degree of adaptation of the collateral vessels. That means, although tissue perfusion is adequate the collateral flow may not increase when the oxygen demand increases because of still high collateral resistance. The measurement of the PCP may help here. We tried, however, a different approach. In those dogs where high basal anastomoses were found by in vivo arteriography a double-tracer technique was used to separate normal coronary flow from collateral flow. When Xenon is injected into the anterior descending artery and Krypton into the occluded circumflex artery, the Krypton-wash out measures only the collateral blood flow. The Xenon-clearance, however, measures the total area since Xenon will enter the occluded artery via the collaterals. When a coronary vasodilating drug of sufficient intensity and duration is injected (SCHAPER et al., 1966), the increase in collateral flow can be compared to the increase in coronary flow. Should the collateral flow fall behind the coronary flow, a second exponential would be expected to appear in the Xenon wash-out and a less substantial increase is expected in the Krypton-clearance; that is, the first exponentials of both Xenon and Krypton are also expected to differ.

This is indeed the case in dogs a few days after coronary artery occlusion (or three weeks after the implantation of the constrictor). Identity of perfusion was almost reached about 8 weeks after the occlusion. The appearance of a second exponential in the clearance curve after drug-induced coronary vaso-dilation might even prove useful in the clinic whenever radio-Xenon is injected via a Sones-catheter for diagnostic purposes. Clinical cardiologists are often disappointed (BALTAXE et al., 1969) when they discover that patients with angina pectoris show the same myocardial clearance of Xenon as normal patients. Unequal distribution of flow within the myocardium can be greatly increased, however, and hence would be detected by graphical analysis of the clearance curve when a vasodilator drug is injected.

IV. The Functional Anatomy of the Collateral System

Several publications from this laboratory (SCHAPER and VANDESTEENE, 1967; SCHAPER et al., 1968, 1969, 1970; BORGERS et al., 1970) have furnished evidence that the enlargement of collaterals and anastomoses is caused by growth, i.e. mitotic cell division of all normal constituents of the arteriolar wall. This was demonstrated by the chance finding of smooth muscle cells in mitosis, by systematically counting mitotic figures after the injection of mitotic blockers such as colchicine and Colcemid[3] (EIGSTI and DUSTIN, 1955) and by the autoradiographic demonstration of radio-active DNA in nuclei after the injection of the labelled precursor H^3-thymidine.

---

3 Demecolcine.

We could show that the mitotic index as well as the labelling index were highest at the moment ($\pm$ a few days) of complete coronary artery occlusion. DNA-synthesis and mitosis in the walls of collaterals decreased gradually thereafter but were still elevated 8 weeks after the operation. One year after the implantation mitoses or labelled nuclei were not observed. Labelling and mitosis were also never observed in normal coronary arteries of dogs and pigs.

The transformation of a small coronary arteriole into a small coronary artery passes through several stages which we have described in detail elsewhere (SCHAPER et al., 1969). Briefly, we found the following typical stages in the dog heart: hypoxic vasodilation is followed by extreme stretching of the vessel wall. Possibly mediated through the action of a greatly increased wall stress the extremely stretched blood vessel is damaged. This damage consists of leakage of plasma proteins into the vascular wall and into the perivascular space. This is followed by an invasion of leucocytes (PMN, monocytes) into the wall and into the perivascular space. A perivascular inflammation, almost an arteriitis when extreme, is the result. Simultaneously with the inflammation DNA-synthesis and mitotic activity begins. The vascular damage is thereby repaired and the vessel grows by a mechanism of "overshoot-repair" (SCHAPER et al., 1969). The adaptation is furthermore characterized by rupture of the internal elastic membrane, development of a subintimal layer of longitudinal smooth muscle and finally the synthesis of a new elastic lamina. After one year the vessels looks almost normal. Although it is admittedly difficult to determine the number of vessels actively participating in the growth-transformation we tried to count all enlarged vessels at different time intervals from arteriograms (SCHAPER et al., 1969). It became apparent that many vessels initially participated but only a few matured into larger arteries in dogs. A sizeable proportion regressed again. This was also confirmed by histological techniques.

These stages can also be observed in the pig although with some deviations. The vessels are located subendocardially in the pig which contrasts with the subepicardial location in the dog heart (SCHAPER et al., 1967). In the pig heart many more vessels participate in the transformation than in the dog heart, even capillaries seem to transform in the pig heart. Preliminary results (SCHAPER and SCHOENMACKERS, to be published) in the human heart revealed that the human heart resembles both the canine and the porcine heart.

1. Factors which Govern the Growth Transformation of Coronary Collaterals

LIEBOW (1963) in his Handbook article on "Situations which lead to changes in vascular pattern" listed three major factors responsible for vascular transformation: mechanical forces, chemical influences and hereditary factors.

9*

Aside from the hereditary influences which are more or less covered by above described species differences we have studied 3 factors which we feel are of importance. These factors are: tangential wall stress, hypoxia and perivascular inflammation.

2. Tangential Wall Stress and other Physical Forces

The discussion about mechanical influences on the vascular wall as molding forces traces back to Thoma (1893) who stated in his "Laws of Histiodynamics" that

"the diameter of a vessel is determined by the velocity of its blood flow" and that

"the thickness of its wall depends on the tangential wall stress."

Thoma who based his "Laws" on earlier ideas of v. Recklinhausen (1883) was subsequently supported by Nothnagel (1889), by Schoop (1961, 1963) and by Sewell (1967). Since the pressure gradient is closely related to the velocity of the blood stream, the defenders of the pressure gradient theory (Gregg and Fisher, 1963; Eckstein et al., 1941) belong to the followers of Thoma (1893).

We are, however, unable to support the pressure gradient theory for the following reasons: as shown earlier by us (Schaper and Vandesteene, 1967) the collateral vessels expand by growth, i.e. cell division, of all normal cellular constituents of the vascular wall. The physical force influencing this growth-process must therefore act on, say, a medial smooth muscle cell, which subsequently prepares for cell division. The pressure gradient is not such a force which influences a single cell because of the dimension of this force.

The pressure gradient is the pressure difference per unit vessel length. When the length is reduced to the dimensions of one single cell the difference becomes so incredibly small that the only force which can possibly act on the cell is the actual intravascular pressure at that particular point. But even this pressure can not act directly on the cell, this can happen only via the tangential wall tension

$$T = \frac{P \cdot R}{\varDelta}$$

which is proportional to the intravascular pressure (P) at a given point, proportional to the radius of the vessel (R) and inversely proportional to the thickness of the vessel wall ($\varDelta$). This is the so-called Law of Laplace in the modification by O. Frank (1920), which had been discussed as the primary molding force in the development of collateral blood vessels by Schaper (1967). There are, however, different ways of thinking to show the validity of our hypothesis on the causal chain of events (i.e. hypoxic vasodilation → increased wall stress → reduced resistance by vascular growth → changes in the pressure gradient, velocity of flow and back pressure) (see network theory).

The passive role of the pressure gradient is further amplified by observations regarding the actions of the extravascular resistance. It is well known since the pioneering work of GREGG (1950) that the coronary arterial inflow abruptly falls with the myocardial contraction. This has been named the extravascular component of resistance to flow. It is well known, at least since 1950, that the back pressure rises and falls with the aortic pressure with only minor differences in form and timing. An analysis of this phenomenon was offered in 1967 by SCHAPER: with each cardiac contraction the extravascular resistance rises, the flow over the collateral falls and hence flow-dependent pressure drop over the collateral is diminished. As a consequence the difference between the aortic and the back pressure is also diminished, i.e. the systolic back pressure rises, absolutely and relatively.

Fig. 7 illustrates this behavior. The ratio between systolic pairs of pressures (systolic back pressure divided by systolic aortic pressure) is always higher than the ratio of the diastolic pressures. Fig. 7 shows quite clearly that also the vascular (collateral) resistance influences the difference between the systolic and diastolic ratios. During the phase of rapid increase of the back pressure the lumped collateral resistance has not yet decreased to its lowest possible value, hence the collateral flow during diastole causes a relatively large pressure drop and the difference between systolic and diastolic ratios is greater (0.3) than half a year after the operation when this difference has decreased to 0.05.

I believe firmly in the role of the tangential wall stress for several reasons: firstly, I have shown (SCHAPER, 1967) that the tangential wall stress is increased in the early phases of collateral enlargement and that it is a function of the wall thickness. Secondly, pressure changes exert their influence on the vascular wall only as tension changes.

Wall tension can, however, not explain everything. Although I have shown that the concept of wall stress can be applied and that it follows the characteristics of a self-inhibiting system (i.e. it can explain expansion and it can explain when and why the system comes to a standstill), I could not show a cause-and-effect relationship, but rather a parallelity of events. But this applies for every physical force.

One is especially at a loss when it comes to the problem how a physical force, which acts mainly on the cell membrane, can set off the mechanisms of cellular proliferation which are located in the nucleus. This means: a direct influence of polar stress would call for the reversal of the normal sequence of events which are believed to start in the nucleus.

When consulting the literature, however, I found that membranous and cytoplasmatic influences are known to produce signals which trigger the nuclear events. In 1969, NOEL DE TERRA published a review with the suggestive title "Cytoplasmatic Control over the Nuclear Events of Cell Reproduction", which shows clearly that such mechanisms are operative in primitive organisms such as *Stentor* and *Tetrahymena*.

On the other hand, the force opposing the velocity of blood through arteries, i.e. the "viscous drag" at the blood vascular wall interface may not be neglected. Fry (1968) has shown that endothelial damage can result above a critical velocity. In my opinion a certain minimum velocity is necessary to ensure structural integrity of a transformed vessel: decrease in flow leading to stagnation, as found in smaller collaterals in the vicinity of larger ones, certainly acts as a stimulus to degeneration which was described by us recently (Schaper et al., 1969).

3. Hypoxia

Since the growth of collaterals is closely associated with hypoxia the idea of a growth-promoting metabolic intermediate substance is indeed highly suggestive. By analogy with the hypothesis of the "metabolic autoregulation" of coronary blood flow one might propose a substance which appears during hypoxia in the extracellular space and which has the power to switch on the generative cell cycle of smooth muscle cells, i.e. activates DNA-polymerase etc. If, however, one is probing deeper into the problems of cellular proliferation and cellular differentiation such a substance with a positive "switching" action becomes highly improbable. Tissue culture experiments teach us that in fact the order of events may be reversed. If one explants an artery into a suitable medium, all its cellular constituents will almost immediately start to proliferate. This suggests that in the intact organ the artery must have been under the influence of a growth-repressing factor, which had inactivated the natural tendency of almost every cell to proliferate. This is, under physiological conditions, a very useful mechanism which ensures normal function and structure. Its absence is said to be one of the possible causes of cancer. Such inhibitors of growth have been found in various organs such as in the skin, i.e. chalones, water soluble proteins which stop the proliferative cycle during G_2 (Bullough, 1969). Growth inhibitors have also been found in ultrafiltrates of chick embryo extract (Coogan et al., 1969). It is thus very likely that hypoxia inactivates an inhibitor of growth to which smooth muscle cells and endothelium respond with proliferation. It is interesting to note that Florentin et al., (1969) have proposed such a mechanism to explain endothelial proliferation after short-term cholesterol feeding in pigs. They believe (personal communication) that cholesterol might inhibit an endothelial growth inhibitor of a probably histone-like structure.

On the other hand, the existence of a positive growth-promoting factor should not be rejected altogether. A nerve growth factor was isolated from the submandibular gland (Scott and Liu, 1964). It is, however, difficult to see how such a substance can exert a local effect when the place of its production is at a distance from the place of its activity. Already in 1926, local factors governing hypertrophy of skeletal muscle were assumed to exist by SIEBERT and PETROW and growth hormone was shown to stimulate reparative changes in anoxic heart muscle by GUDBJARNASON et al. (1966).

A very attractive hypothesis would be the following chain of events: during hypoxia, or under the influence of extracellular metabolites from hypoxic heart muscle, the metabolic pathway of vascular smooth muscle switches from the Krebs-cycle to the Pentose Phosphate Pathway (PPP) which is also called hexosemonophosphate shunt. This has several advantages. Firstly, it necessitates a smaller "investment" of ATP for the enzymatic reactions and, secondly, it provides the cells with the precursors of the nucleotides necessary for cellular proliferation.

PPP has been shown to be activated in infarcted myocardium by GUDBJANASON et al., (1968) and BRAASCH (1955) but no information is at present available whether this had taken place in mesenchymal or myocardial tissue.

4. Perivascular Inflammation (P.I.)

Inflammation per se is one of the most powerful situations known to produce blood vessels, mainly capillaries. The process of budding and sprouting of capillaries was elegantly demonstrated by the CLARKS (1932) many years ago. We have seen furthermore (SCHAPER, unpublished observations) that, in the rat, arteries in the vicinity of experimental cotton-pellet granulomas grow larger. This was shown by the autoradiographic demonstration of radio-active DNA after injection of H^3-thymidine. It is known that degenerating leucocytes in inflamed tissue are powerful stimulators of cellular proliferation and the ultrafiltered medium of leucocytes in cell culture contains mitogenic factors (IMRIE, 1968). Viewed from this angle the observation of perivascular inflammation is indeed intriguing because its very presence suggests a potentiation of the growth transformation of collaterals. The P.I. is thought to be caused as follows: stretching of and damage to the vascular wall favours the insudation of plasma-proteins and of leucocytes into the media and into the perivascular space. The presence of these proteins in the perivascular space plus hypoxia of the surrounding cardiac muscle increases the permeability of the venules and more leucocytes invade the perivascular space which results in P.I. This either triggers or potentiates cell proliferation.

5. The Concerted Action of Wall Stress, Hypoxia and P.I.

When the process of growth transformation starts all 3 factors are present and it is difficult if not impossible to estimate the role played by one of these factors. Later on, however, these influences can be separated.

Several days after the coronary artery occlusion hypoxia ceases to be a dominating factor in dogs because the vessels have enlarged so much that adequate perfusion is guaranteed. P.I., however, is still present and may be operative in vascular transformation. Several weeks after occlusion P.I. ceases to be an important factor but wall stress is still operative and it acts to increase the thickness of the wall.

The interplay of these factors is also modulated by heredity. In the dog the above described time dissociation is quite evident. In the pig heart several modifications are known. The foreign body type of inflammation which exists around the implanted coronary artery constrictor is a surprisingly efficient source of collaterals in the pig heart but not in the dog. P.I. is observed in the pig heart but to a lesser degree than in the dog. Tangential wall stress is very high in newly formed pig collaterals (they are extremely thin-walled!), but the arterial response to it seems weaker in the pig heart (vessels remain thin-walled longer).

6. The Situation in the Human Heart

It has been pointed out in the introduction that collaterals and anastomoses do exist in the normal human heart and it is a well proven and well accepted fact that these vessels enlarge under the influence of localized ischemia. The demonstration of these vessels is to a large part a matter of technique as was beautifully documented by FULTON (1965). Only three methods are at present available for the visualization of anastomoses. These are:

1. the post-mortem injection of radio-opaque material and subsequent arteriography using special dissection and stereo-X-ray techniques.

2. the injected heart is made translucent by chemical means and the anastomoses can be counted and measured from high magnification photographs.

3. the third method is in vivo cine-angiography which rapidly develops into a relatively safe diagnostic procedure in many clinical cardiac catheterization laboratories.

Huge differences exist with regard to the resolving power of these three methods. The "injection-and-clearing" method of SPALTEHOLZ (1907) resolves vessels down to about 10–15 μm provided that the injectate has penetrated properly. Post-mortem arteriography can demonstrate vessels down to about 25 μm provided that fine-grain high resolution X-ray film is used. Cine-coronaryangiography has the lowest resolving power. Demonstrable anastomoses should measure at least 200 μm and they have to run at the epicardial surface of the heart were they are rarely found. Usually the presence of anastomoses is inferred when retrograde filling of occluded arteries on cine-angiographic films is demonstrable. The deep subendocardial plexus of anastomoses which is almost always responsible for the retrograde filling is, however, not visible.

Although there is a gradient of resolution from "clearing"-results to cine-angiography an inverse relation exists as to the efficacy of these vessels.

GREGG (1963) quite correctly pointed out that the post-mortem demonstration of anastomoses does not necessarily imply that these vessels have had a function during life. FULTON (1965) demonstrated without any doubt that ischemic heart disease caused by coronary artery thrombosis leads to an enlargement of preexistent anastomoses. The rather large anastomoses near or within infarcted heart muscle which FULTON so beautifully documented became visible during the injection of the radioopaque mass at physiological pressure (100–150 mm Hg) only after several minutes and sometimes only after hours of continued injection. This was especially the case when the collateral vessels were supplied by an artery which itself developed stenosis after the anastomoses had developed. Although these anastomoses were of large caliber their function as blood conductors was probably rather low. Since cine-angiography is closer to the actual pathophysiological conditions during life even the barely visible collaterals this method is able to show are probably of much greater importance to the muscle they supply with blood. It can be assumed that a vessel which transports contrast material during cine-angiography will also transport blood. No inference, however, can be made from cine-angiography as to the amount of blood this vessel may carry.

The origin of the enlarged collaterals in localized ischemic heart disease is believed by FULTON to stem from preexistent vessels. It is, on the other hand, not entirely clear whether newly formed capillaries can transform into anastomotic arteries and, when they do, what their physiological significance is.

Our own experiments (SCHAPER and colleagues, unpublished) have shown that sprouting capillaries in granulation tissue either in the vicinity of the constricting device (= foreign body reaction) or within an infarct can enlarge substantially. Sometimes these vessels have developed into anastomotic arteries of several hundred micrometers, especially within the granulation tissue around the constrictor. This capillary enlargement and transformation was most prominent in the pig but was also observed in the dog heart. They were demonstrated by arteriography, histologically and by DNA-synthesis studies using H^3Tdr as a precursor of DNA. I believe that capillary enlargement does play a role in the collateralization of the human heart. The extremely dense vascularization of infarcted human heart muscle and of areas near older infarcts looks suspicious for capillary transformation. The structure of these vessels resembles also those of extremely enlarged capillaries since they are lacking a muscular coat.

7. The Histological Structure of Human Anastomoses [4]

Very little is known about the histological structure of human collaterals and we know of no reports about systematic studies in this field. Only oc-

4 The human heart studies were carried out in the department of Pathology, Medical Faculty, Technical University of Aix La Chapelle, Germany. These studies were made possible through the kind cooperation of Prof. J. SCHOENMACKERS, head of above mentioned department, whose help and encouragement is gratefully acknowledged.

casionally mention is made (BAROLDI, 1967) that human anastomoses are extremely thin-walled, that they look like overstretched arterioles or that they are lacking smooth muscle and hence resemble hugely dilated capillaries. FULTON (1965) summarized his observations as follows: "In the histological section, the walls of vessels which were believed to be anastomoses were exceedingly thin, being composed of little more than an endothelial lining, supported on a thin fibrous tunic, in which elastic tissue was scanty and only occasional isolated muscle cells were found. Even when such channels were expanded under the influence of disease to a diameter of 500 microns or more, the wall remained a thin structure as described".

Because of these scanty observations we found it worth while to compare the histological structure of canine and porcine collaterals with those of the human heart.

The histological investigation of human collaterals is much more complicated than the study of canine collaterals. When a coronary artery is slowly constricted experimentally in the dog heart the anastomoses develop mainly on the epicardial surface of the heart and donor artery (= stem), midzone and re-entry into the occluded recipient vessel are easily visible to the naked eye. When such an anastomosis is cut out over its entire length and then divided into several small tissue blocks the histological investigation can be carried out at various well known time intervals after coronary artery occlusion. In the human heart the enlarged anastomoses are usually not visible to the naked eye. Even epicardial connections are mostly burried in the epicardial fat and the dissection of collaterals can only be done when constantly checking their localization on stereo-X-ray films. A further complicating factor is the frequent occurrence of multiple coronary artery disease and it is often difficult to define stem and re-entry because the direction of blood flow cannot always be diagnosed from the arteriogram. Because of these difficulties only midzone-vessels were investigated. Information on about 50 human hearts is available at present. All patients had long-standing atherosclerotic and/or thrombotic coronary artery stenosis or occlusion and they had died either from re-infarction or from causes completely unrelated to coronary artery disease. Because of our previous knowledge derived from a large number of tissue sections from canine and porcine collaterals we looked for those vascular changes which we consider typical of collaterals in experimental animals. These typical stages in animals are in particular the extremely thin-walled enlarged vessels early after coronary artery occlusion and the perivascular inflammatory reaction. These vessels show a more or less normal arterial structure several weeks to months after the occlusion but the fragmented internal elastic membrane is characteristically displaced toward the tunica media.

All these typical features previously found in experimental coronary occlusion in animals were also found in the human heart. In patients dying from acute myocardial infarction, or several days after the infarct, extremely thin-walled vessels were found near the infarcted area and the perivascular inflammatory reaction was always present. This reaction varied from a mild exsudation of plasma protein, which was previously described by KENT (1967), to a very severe inflammatory response including round-cell infiltration of the vascular wall and its tunica adventitia.

In a patient who died from the sequel of silicosis an old isolated thrombotic occlusion of the anterior descending artery was found which was partly re-canalized. Three large epicardial anastomoses were found which exhibited a normal arterial structure with the exception of a displaced probably formerly ruptured internal elastic membrane and a mild subintimal thickening which is also frequently found in dogs one year after coronary artery occlusion. Similar histological pictures were found in the hearts of patients dying from re-infarctions and long-standing collaterals from previous infarctions were investigated. Many times the puzzling observation of an extremely dense vascularization of scar tissue was made. As pointed out above these vessels look suggestive of transformed capillaries. This observation is astonishing because it is well known that capillaries in granulation tissue usually regress when the granulation tissue transforms into scar-tissue. The existence of pressure gradients between perfusion areas may have prevented the regression of these vessels.

Summarizing we can say that the histological structure of human collaterals resembles the vascular changes seen in experimental animals with chronic coronary artery occlusion. Species differences between the vascular reactions of human, canine or porcine coronary vessels could not be detected except for localization. Although epicardial, endomural and subendocardial anastomoses were found in all three species studied, the dog heart exhibits a preference for epicardial and the human and pig heart show a preference for endomural and subendocardial anastomoses.

8. Summary of New Concepts for Future Research

This review of the current literature and of our own experimental work on the physiology of the collateral circulation tried to demonstrate the astonishing adaptive changes which coronary vessels can undergo. These adaptive changes consisted mainly of growth-transformation of a pre-existent network of arteriolar and pre-capillary connections. All cell-types normally found in the vascular wall and its immediate adventitial surroundings, i.e. endothelium, smooth muscle, fibroblasts and other mesenchymal cells, proliferate by mitotic division. The increased mitotic activity is the result of the interaction of biochemical and physical forces which are to a large part the more or less direct consequences of tissue hypoxia. Physical forces, such as increased wall tension, may, however, play a role independent of ischemia.

Experimental studies and clinical observations have shown that coronary artery occlusion does not necessarily produce a myocardial infarct provided that the rate of arterial narrowing is sufficiently slow for the collaterals to develop. It is a well accepted fact that a slight luminal narrowing of a main coronary artery is not a sufficient stimulus for the growth-transformation of anastomoses. The sudden thrombotic occlusion of such a slightly diseased artery strikes therefore a completely unprepared heart muscle and myocardial infarction is the inevitable result. Since, in most cases, slight narrowing is not diagnosed and the prevention of coronary artery thrombosis by chronic anticoagulation therapy years before the expected infarct is not a well accepted clinical way to cope with the problem, other ways of prophylaxis must be developed. Regular physical exercise should be encouraged in such a way that it becomes habit forming during almost the entire life span. Such a program of exercise conditioning has several physiological consequences: the increased oxygen expenditure of the heart during exercise may cause a slight luminal narrowing to behave like a critical stenosis which in turn acts as a stimulus for collateral enlargement. Physical exercise is also expected to increase the size of the interconnecting arteriolar network which consequently becomes better adapted for its later potential role as a source of collaterals. Physical exercise may also cause an enlargement of the main epicardial coronary arteries so that an atherosclerotic plaque in a widened blood conductor becomes less life threatening.

Apart from the direct influences of physical training on the coronary arteries the slowing of the heart rate as a consequence of training is of additional beneficial value because less oxygen is needed by the heart for the same amount of work.

The value of drug therapy with so-called coronary vasodilators has not been fully evaluated at present, therefore the use of these drugs should not be uncritically discouraged. The problems of adequate blood levels, gastrointestinal absorption, correct dose and long term results with these drugs are far from being solved. General and maximal vasodilation of the entire coronary arterial tree may be dangerous because of the possible redistribution of blood flow which can result in a deprivation of oxygen from the area where it is needed most. This depends mainly on the degree of arterial narrowing and on the degree of collateralization. It is, on the other hand, known that drug-induced vasodilation increases the size of the interconnecting arteriolar network in the normal heart and in hearts with one occluded coronary artery. The reduction of platelet-aggregation by some of these drugs offers another interesting possibility of prophylaxis.

More efforts should be made to develop drugs which selectively act on the collateral vessels itself and on the resistance vessels supplied by collaterals. Although this sounds to be very difficult to achieve at least the methodology for testing is available now.

The influence of growth-stimulators, mitogenic agents and hormones (somatotropine, adrenocortical hormones, thyroxine and its derivatives) on the enlargement of collaterals has not received much attention. The biochemical transmitters responsible for the growth of blood vessels in localized ischemia have not been identified so far with conclusive evidence. The causes for collateral regression already anticipated by GREGG (1963) and more and more recognized experimentally recently (SCHAPER et al., 1969) are poorly understood.

A most rewarding terrain for future research lies before us and it should be evaluated with every available method, ranging from the methods of molecular and cell biology over hemodynamic studies to pharmacology and surgery.

References

ANREP, G. V., HÄUSLER, H.: The coronary circulation. I. The effect of changes of the blood pressure and of the output of the heart. J. Physiol. (Lond.) 65, 357—373 (1928).

BALTAXE, H. A., FORMANEK, G., LOKEN, M., AMPLATZ, K.: Clinical limitations to use of Xenon for measurement of myocardial blood flow. Invest. Radiol. 4, 317–322 (1969).

BAROLDI, G., MANTERO, O., SCOMAZZONI, G.: The collaterals of the coronary arteries in normal and pathologic hearts. Circulat. Res. 4, 223–229 (1956).

— SCOMAZZONI, G.: Coronary circulation in the normal and the pathologic heart. Office of the surgeon general Department of the Army. Washington, D. C. 1967.

BECKER, L., FORTUIN, N. J., PITT, B.: Regional myocardial blood flow in the conscious dog. Circulation 40, III 41 (1969).

BELLMAN, S., FRANK, H. A.: Intercoronary collaterals in normal hearts. J. thorac. Surg. 36, 584–603 (1958).

BERNE, R. M., JONES, R. D., CROSS, F. S.: Evaluation of procedures designed to enhance collateral blood flow. Circulat. Res. 10, 142–147 (1962).

BETZ, E., KRUG, A., SCHMAHL, F. W.: Die lokale Myokarddurchblutung bei Ligatur von Koronargefäßen. Verh. Dtsch. Ges. Kreisl.-Forsch., 30. Tagg. p. 273–279 (1964).

BLAIR, E.: Anatomy of the ventricular coronary arteries in the dog. Circulat. Res. 9, 333–341 (1961).

BLOOR, C. M., ROBERTS, L. E.: Effect of intravascular isotope content on the isotopic determination of coronary collateral blood flow. Circulat. Res. 26, 537–544 (1965).

BLUMGART, H. L., ZOLL, P. M., FREEDBERG, A. S., GILLIGAN, D. R.: The experimental production of intercoronary arterial anastomoses and their functional significance. Circulation 1, 10–27 (1950).

BORGERS, M., SCHAPER, J., SCHAPER, W.: Acute vascular lesions in developing coronary collaterals. Virchows Arch. Abt. A Path. Anat. 351, 1–11 (1970).

BRAASCH, W.: Veränderungen des Herzstoffwechsels während der Infarktheilung. Arzneimittel-Forsch. 5, 799–802 (1955).

BULLOUGH, W. S.: The chalones. Sci. J. 5, 71–76 (1969).

BURCHELL, H. B.: Adjustments in coronary circulation after experimental coronary occlusion with particular reference to vascularization of pericardial adhesion. Arch. intern. Med. 65, 240–262 (1940).

BURGISON, R. M., GORDON, G. L., ADAMS, C. R., KRANTZ, J. C.: Studies on a new coronary vasodilator, 1-chloro-2, 3-propanediol dinitrate. Angiology 13, 412–417 (1962).

CANNON, P. J., HAFT, J. I., JOHNSON, P. M.: Visual assessment of regional myocardial perfusion utilizing radio-active Xenon and scintillation photography. Circulation 15, 277–288 (1969).

Clark, E. R., Clark, E. L.: Observations on living preformed vessels as seen in a transparent chamber inserted in the rabbit ear. Amer. J. Anat. **49**, 441–447 (1932).

Coogan, P. S., Friede, J., Maganini, H., Berardi, R. S., Hass, G. M.: Purification of growth inhibitors in ultrafiltrate of chick embryo extract. Lab. Invest. **20**, 371–376 (1969).

Davidson, A. I. G., Leighninger, D. S.: Collateral flow in the heart at different aortic perfusion pressures using left heart bypass: an experimental study. Ann. Surg. **162**, 48–52 (1965).

De Terra, N.: Cytoplasmatic control over the nuclear events of cell reproduction. Int. Rev. Cytol. **25**, 1–29 (1969).

Domenech, R. J., Hoffman, J. I. E., Noble, M. I. M., Saunders, K. B., Subijanto, S.: Total and regional coronary blood flow measured by radio-active microspheres in conscious and anesthetized dogs. Circulat. Res. **25**, 581–596 (1969).

Doutheil, U., Rohde, R.: Messung der Koronadurchblutung mit der Auswaschtechnik radioaktiver inerter Gase (Xe-133, Kr-85) im Zusammenhang mit der Frage nach der Durchblutungsverteilung im Myokard. Verh. dtsch. Ges. Kreisl.-Forsch. **32**, 273–278 (1966).

Eckstein, R. W.: Coronary interarterial anastomoses in young pigs and mongrel dogs. Circulat. Res. **2**, 460–465 (1954).

— Development of interarterial coronary anastomoses by chronic anemia. Disappearance following correction of anemia. Circulat. Res. **3**, 306–310 (1955).

— Effect of exercise and coronary artery narrowing on coronary collateral circulation. Circulat. Res. **5**, 230–235 (1957).

— Gregg, D. E., Pritchard, W. H.: The magnitude and time of development of the collateral circulation in occluded femoral, carotid and coronary arteries. Amer. J. Physiol. **132**, 351–361 (1941).

Eigsti, O. J., Dustin, P.: Colchicine. Iowa: Iowa State Univ. Press 1955.

Fam, W. M., McGregor, M.: Effect of coronary vasodilator drugs on retrograde flow in areas of chronic myocardial ischemia. Circulat. Res. **15**, 355–365 (1964).

Florentin, R. A., Nam, S. C., Lee, K. T., Thomas, W. A.: Increased ^{3}H-Thymidine incorporation into endothelial cells of swine fed cholesterol for three days. Exp. molec. Path. **10**, 250–255 (1969).

Frank, O.: Die Elastizität der Blutgefäße. Z. Biol. **71**, 253–272 (1920).

Fry, D. L.: Acute endothelial changes associated with increased blood velocity gradients. Circulat. Res. **22**, 165–197 (1968).

Fulton, W. F. M.: The coronary arteries. Springfield, Ill.: Ch. C. Thomas 1965.

Giese, W., Müller-Mohnssen, H.: Kollateralkreisläufe im Coronarsystem bei Coronarsklerose. Bad Oeynhausener Gespräche II, p. 157–177 (1958).

Gömöri, Z.: Beitrag zum postmortalen Nachweis interarterieller koronarer Anastomosen im menschlichen Herzen. Z. Kreisl.-Forsch. **54**, 1181–1189 (1965).

Grayson, J., Irvine, M., Parratt, J. R., Cunningham, J.: Vasospasic elements in myocardial infarction following coronary occlusion in the dog. Cardiovasc. Res. **2**, 54–62 (1968).

Gregg, D. E.: Coronary circulation in health and disease. Philadelphia: Lea & Febiger 1950.

— Fischer, L. C.: Blood supply to the heart. Handbook of physiology, vol. 2, sect. 2, p. 1517–1584. Baltimore: Williams & Wilkins 1963.

— Thornton, J. J., Mautz, F. R.: The magnitude, adequacy and source of the collateral blood flow and pressure in chronically occluded coronary arteries. Amer. J. Physiol. **127**, 161–175 (1939).

Gudbjarnason, S., Braasch, W., Cowan, C., Bing, R. J.: Metabolism of infarcted heart muscle during tissue repair. Amer. J. Cardiol. **22**, 360–369 (1968).

— Fenton, J. C., Wolf, P. L., Bing, R. J.: Stimulation of reparative processes following experimental myocardial infarction. Arch. intern. Med. **118**, 33–40 (1966).

HAFT, J. I., DAMATO, A. N.: Measurement of collateral blood flow after myocardial infarction in the closed-chest dog. Amer. Heart J. **77**, 641–648 (1969).

HARMAN, M. A., MARKOV, A., LEHAN, P. H., OLDEWURTEL, H. A., REGAN, T. J.: Coronary blood flow measurements in the presence of arterial obstruction. Circulat. Res. **19**, 632–637 (1966).

HERD, J. A., BARGER, A. C.: Simplified technique for chronic catheterization of blood vessels. J. appl. Physiol. **19**, 791–792 (1964).

HUSAK, V., PERINOVA, U.: The design of collimators for radioisotope scanning. Phys. in Med. Biol. **14**, 233–244 (1969).

IMRIE, R. C., MÜLLER, G. C.: Release of lymphocyte growth promotor in leucocyte cultures. Nature (Lond.) **219**, 1277–1279 (1968).

JAMES, T. N.: Anatomy of the coronary arteries. New York: Paul B. Hoeber, Inc. 1961.

JOHANSSON, B., LINDER, E., SEEMAN, T.: Collateral blood flow in the myocardium of dogs measured with Krypton-85. Acta physiol. scand. **62**, 263–270 (1964).

— — — Coronary collateral blood flow in relation to the mass of ischemic myocardium, studied with Krypton-85. Acta physiol. scand. **63**, 495–504 (1965).

— — — Effects of hematocrit and blood viscosity on myocardial blood flow during temporary coronary occlusion in dogs. Scand. J. thorac. cardiovasc. Surg. **1**, 165–174 (1967).

JOHNSON, J. R., DI PALMA, J. R.: Intramyocardial pressure and its relation to aortic blood pressure. Amer. J. Physiol. **125**, 234–240 (1939).

KATTUS, A. A., GREGG, D. E.: Some determinants of coronary collateral blood flow in the open-chest dog. Circulat. Res. **7**, 628–642 (1959).

KENT, S. P.: Diffusion of plasma proteins into cells: a manifestation of cell injury in human myocardial ischemia. Amer. J. Path. **50**, 623–637 (1967).

KIRK, E. S., HONIG, C. R.: An experimental and theoretical analysis of the myocardial tissue pressure. Amer. J. Physiol. **207**, 361–367 (1964).

KREUZER, H., SCHOEPPE, W.: Das Verhalten des Druckes in der Herzwand. Pflügers Arch. ges. Physiol. **278**, 181–198 (1963).

LANGE, G.: Messung der Durchblutung des experimentellen Herzinfarkts und Beeinflussung der Durchblutung des infarzierten und nicht infarzierten Herzmuskels durch einige Gifte. Naunyn-Schmiedebergs Arch. exp. Path. Pharmak. **246**, 240–286 (1963).

LAURIE, W., WOODS, J. D.: Anastomoses of the coronary circulation. Lancet **1958 II**, 812–816.

LEIGHNINGER, D. S., EINSEL, I. H., RUEGER, R. G., BECK, C. S.: Intercoronary arterial channels produced by chemical agents. Amer. J. Cardiol. **6**, 949–951 (1960).

LEVY, M. N., CHANSKY, M.: Collateral circulation after coronary artery constriction. Amer. J. Physiol. **208**, 144–148 (1965).

— IMPERIAL, E. S., ZIESKE, H.: Collateral blood flow to the myocardium as determined by the clearance of rubidium-86 chloride. Circulat. Res. **9**, 1035–1043 (1961).

LIEBOW, A. A.: Situations which lead to changes in vascular pattern. Handbook of physiology, vol. 2, sect. 2, p. 1251–1276. Baltimore: Williams & Wilkins 1963.

LINDER, E.: Measurements of normal and collateral coronary blood flow in dogs by Krypton-85 and Xenon-133. Acta physiol. scand. **68**, Suppl. 272, 5–31 (1966).

— Studies on coronary collateral blood flow in dogs by Krypton-85 and Xenon-133 clearance. Göteborg 1966.

— SEEMAN, T.: Effects of persantin and nitroglycerin on myocardial blood flow during temporary coronary occlusion in dogs. Angiologica **4**, 225–255 (1967).

LITVAK, J., SIDERIDES, L. E., VINEBERG, A. M.: The experimental production of coronary artery insufficiency and occlusion. Amer. Heart J. **53**, 505–518 (1957).

LOWER, R.: Tractus de Corde. Amsterdam: Elsevier 1669.

LUMB, G. D., HARDY, L. B.: Collateral circulation and survival related to gradual occlusion of the right coronary artery in the pig. Circulation **27**, 717–721 (1963).

Lumb, G., Singletary, H. P., Hardy, L. B.: Acute and gradual coronary occlusion in the pig. Fed. Proc. **21**, 98 (1962).

McLean, L. D., Hedenstrom, P. H., Rayner, R. R.: Tissue blood flow to the heart. Influence of coronary occlusion and surgical measures. Circulat. Res. **10**, 45–50 (1962).

Moore, R. A.: The coronary arteries of the dog. Amer. Heart J. **5**, 743–755 (1929/30).

Neutze, J. M., Wyler, F., Rudolph, A. M.: Use of radioactive microspheres to assess distribution of cardiac output in rabbits. Amer. J. Physiol. **215**, 486–495 (1968).

Nothnagel, H.: Über Anpassungen und Ausgleichungen bei pathologischen Zuständen. III. Abhandlung: Die Entstehung des Kollateralkreislaufs. Z. klin. Med. **15**, 42–55 (1889).

Pap, J., Török, B., Toth, I., Temes, G., Bartos, G., Kustos, G.: Die Beurteilung des „Rückfluß"-Testes nach akuter Okklusion eines Hauptkoronarastes. Thoraxchir. Vask. Chir. **12**, 373–376 (1965).

Pianetto, M. B.: The coronary arteries of the dog. Amer. Heart J. **18**, 403–410 (1939).

Pitt, B.: Interarterial coronary anastomoses. Occurence in normal hearts and in certain pathological conditions. Circulation **20**, 816–822 (1959).

Prinzmetal, M., Simkin, B., Bergmann, H. C., Kruger, H. E.: Studies on the coronary circulation. II. The collateral circulation of the normal human heart by coronary perfusion with radioactive erythrocytes and glass spheres. Amer. Heart J. **33**, 420–442 (1947).

Ravin, A., Geever, E. F.: Coronary arteriosclerosis, coronary anastomoses and myocardial infarction. Arch. intern. Med. **78**, 125–138 (1946).

Recklinhausen, F. D. v.: Handbuch der allgemeinen Pathologie des Kreislaufs und der Ernährung. Stuttgart 1883.

Rees, R. J., Redding, V. J.: Anastomotic blood flow in experimental myocardial infarction. A new method using Xenon-133 clearance. Cardiovasc. Res. **1**, 169–178 (1967).

— — Effects of Dipyridamole on anastomotic blood flow in experimental myocardial infarction. Cardiovasc. Res. **1**, 179–183 (1967).

— — Experimental myocardial infarction by a wedge method: early changes in collateral flow. Cardiovasc. Res. **2**, 43–53 (1968).

— — Experimental myocardial infarction in the dog: comparison of myocardial blood flow within, near, and distant from the infarct. Circulat. Res. **25**, 161–170 (1969).

— — Increase in myocardial collateral capacity following drug induced coronary vasodilation. Amer. Heart J. **78**, 224–228 (1969).

— — Ashfield, R., Gibson, D., Garey, C. J.: Myocardial blood flow measured with Xenon-133. Effect of glyceryl trinitrate in dogs. Brit. Heart J. **28**, 374–381 (1966).

Reiner, L., Molnar, J., Jimenez, F. A., Freudenthal, R. R.: Interarterial coronary anastomoses in neonates. Arch. Path. **71**, 103–112 (1961).

Rudolph, A. M., Heymann, M. A.: The circulation of the fetus in utero. Circulat. Res. **21**, 163–184 (1967).

Schaper, J., Borgers, M., Schaper, W.: Ultrastructure and histochemistry of developing intercoronary anastomoses. [Proc. of the 5th European Congr. of Cardiology, p. 321–326 (1968).

— — Coronary collateral vessels. An electronmicroscopic study. Exp. molec. Path· (1970) (in press).

Schaper, W.: Der Einfluß physikalischer Faktoren auf das Radialwachstum von Kollateralgefäßen im Koronarkreislauf. Verh. Dtsch. Ges. Kreisl.-Forsch. **32**. Tagg, p. 282–286 (1966).

— Tangential wall stress as a molding force in the development of collateral vessels in the canine heart. Experientia (Basel) **23**, 595 (1967).

— The collateral circulation in the canine coronary system. Thesis, Univ. of Louvain, 1967.

— Collateral circulation after experimental coronary artery occlusion. Acta cardiol. (Brux.), Suppl. **13**, 74 (1969).

SCHAPER, W., JAGENEAU, A., XHONNEUX, R.: The development of a collateral circulation in the pig and dog heart. Cardiologia (Basel) 51, 321–335 (1967).
— SCHAPER, J., XHONNEUX, R., VANDESTEENE, R.: The morphology of intercoronary anastomoses in chronic coronary artery occlusion. Cardiovasc. Res. 3, 315–323 (1969).
— VANDESTEENE, R.: The rate of growth of interarterial anastomoses in chronic coronary artery occlusion. Life Sci. 6, 1673–1680 (1967).
SCHAPER, W. K. A., XHONNEUX, R., JAGENEAU, A. H. M., JANSSEN, P. A. J.: The cardiovascular pharmacology of Lidoflazine, a long-acting coronary vasodilator. J. Pharmacol. exp. Ther. 152, 265–274 (1966).
SCHLESINGER, M. J.: An injection plus dissection study of coronary artery occlusions and anastomoses. Amer. Heart J. 15, 528–568 (1938).
SCHMIDT, H. D., SCHMIER, J.: Erhöhte Toleranz gegen Coronarverschluß durch ein pharmakologisch vermehrtes Gefäßnetz. Arzneimittel-Forsch. 16, 1058–1064 (1966).
SCHOENMACKERS, J.: Zur Anatomie und Pathologie der Coronargefäße. Bad Oeynhausener Gespräche II, p. 133–156 (1958).
SCHOOP, W.: Pathophysiologie und Klinik des arteriellen Kolateralkreislaufs beim Verschluß von Extremitätenarterien. Habil.-Schr. Univ. Freiburg i. Br., 1963.
— JAHN, W.: Entwicklungsstadien arterieller Kollateralen und ihre begriffliche Definition. Z. Kreisl.-Forsch. 50, 249–261 (1961).
SCOTT, D., LIU, C. N.: Factors promoting regeneration of spinal neurons: positive influence of nerve growth factor. Progress in brain research, vol. 13, ed. M. SINGER and J. P. SCHADÉ; Mechanisms of neural regeneration. Amsterdam: Elsevier 1964.
SEWELL, W. H.: Surgery for acquired coronary disease. Springfield, Ill.: Ch. C. Thomas 1967.
SIEBERT, W., PETOW, H.: Studien über Arbeitshypertrophie des Muskels. Z. klin. Med. 102, 427–434 (1926).
— — Über wachstumsfördernde Substanzen im arbeitenden Muskel. Z. klin. Med. 102, 434–442 (1926).
SPALTEHOLZ, W.: Die Koronararterien des Herzens. Verh. anat. Ges. 21, 141–162 (1907).
— Die Arterien der Herzwand. Leipzig: Hirzel 1924.
THOMA, R.: Untersuchungen über die Histogenese und Histomechanik des Gefäßsystems. Stuttgart: Enke 1893.
VASTESAEGER, M. M., STRAETEN, P. P. VAN DER., FRIART, J., CANDAELE, G., GHYS, A., BERNARD, M.: Les anastomoses intercoronariennes telles qu'elles apparaissent à la coronarographie postmortem. Acta cardiol. (Brux.) 12, 365–401 (1957).
VINEBERG, A., MAHANTI, B., LITVAK, J.: Experimental gradual coronary artery constriction by Ameroid constrictors. Surgery 47, 765–771 (1960).
WETTERER, E., PIEPER, H. P.: Über die Gesamtelastizität des arteriellen Windkessels und ein experimentelles Verfahren zu ihrer Bestimmung am lebenden Tier. Z. Biol. 106, 23–57 (1953).
WIGGERS, C. J.: The problem of functional coronary collaterals. Exp. Med. Surg. 8, 402–421 (1950).
— GREEN, H. D.: The ineffectiveness of drugs upon collateral flow after experimental coronary occlusion in dogs. Amer. Heart J. 11, 527–541 (1936).
ZOLL, P. M., NORMAN, L. R.: The effects of vasomotor drugs and of anemia upon interarterial coronary anastomoses. Circulation 6, 832–842 (1952).
— WESSLER, S., SCHLESINGER, M. J.: Interarterial coronary anastomoses in the human heart with particular reference to anemia and relative cardiac hypoxia. Circulation 4, 797–815 (1951).

Rheological Properties of Human Erythrocytes and their Influence upon the "Anomalous" Viscosity of Blood

H. Schmid-Schönbein* and R. E. Wells, Jr.

With 36 Figures

Table of Contents

* Supported in part by a research fellowship of the "Max Kade Foundation", New York, N.Y.
The work was also supported by grants from the J. A. Hartford Foundation, Inc. and P. H. S. Grant 5-PO1-HE-11306 from N. I. H.

Acknowledgments: The able technical assistance of Mrs. L. Doris, Mr. D. Reilly and Mr. D. Fritz, (Boston) as well as of Mrs. A. Rubia, Miss M. Albrecht and Mrs. U. Kottmayr is gratefully acknowledged.

I. Introduction and History of the Problem

Rheology is the science concerned with the flow and deformation of matter. Its application to the study of the flow properties of blood and blood cellular elements was based on the reasoning that the fluiddynamic behavior of these materials could not be adequately described using the methods of classical hydrodynamics nor the theory of elasticity. The emphasis of such studies in hemorheology has recently been extended beyond the primary hemodynamics of blood viscosity and beyond coagulation as an hemodynamic incident. During the recent years, a more general study of the variables allowing or restricting the ability to flow (fluidity) of cell suspensions in the vasculature has been initiated.

A critical evaluation of these contemporary trends requires a short historical treatise, followed by a critique of the methods and their applicability to the problems of hemorheology. Historically, the suspension characteristics of human blood played an important role in clinical medicine long before their influence on blood viscosity was discovered in the 19th century. As described by FAHRAEUS [51], the therapeutic rationale of phlebotomy was the effect this procedure had upon the rheological properties of blood. Whenever red cells were pathologically aggregated, the cells settled from the plasma before coagulation occured. This was taken as evidence of the dyscrasia of the blood and as the justification for therapeutic phlebotomy. The medieval practice of characterizing disease on the basis of the physical characteristics of blood sedimentation has been continued into the 20th century by the test of the sedi-mentation rate of anticoagulated blood.

The viscous characteristics of red cell suspensions were first examined experimentally approximately 80 years ago. Poiseuille [123] in his classic work on viscosity and flow of fluids, was forced to restrict himself to measure-ments with water. Upon inspection, he observed, however, that the flow behavior of blood in narrow glass tubes deviated from that of an ordinary fluid, in that a layer of cell-poor plasma was formed near the wall of narrow glass tubes (cited by Bayliss [5]).

The influence of the red cell concentration on blood viscosity was not firmly established before the turn of the 20th century [81, 124]. As early as 1915, Hess [82] described the dependence of apparent blood viscosity upon the conditions of measurement. When lowering the perfusion pressure he found a more than proportional decrease in flow rate, a phenomenon soon confirmed by others [120, 133]. To explain this increase in apparent viscosity, Hess postulated the "occurrence of an elastic resistance against deformation within the medium". As the cause of this behavior, he discussed the "pronounced capability of the red cells to attach to one another and to join in long elastic chains". Hess [82] came to the conclusion that blood viscosity should be measured under high driving pressure (more than 100 mm Hg) in order to insure "normal behavior of blood". This suggestion was widely accepted and therefore research into the causes and consequences of the anomality was deferred for another three decades.

During the following years, a great number of factors were explored which had an influence upon the bulk viscosity of blood. These were summarized in 1929 by Neuschloss [120]. At the same time, in vivo observations of red cell deformation during flow in the microcirculation, as well as the phenomenon of cell migration and plasma skimming, were firmly established by Krogh [96]. Fahraeus [51, 52] rediscovered the phenomenon of red cell aggregation in disease and its correlation with "granular flow" in the microcirculation in vivo [122], i.e. the observation of flowing aggregates in the microcirculation of the human conjunctiva and the retinal vessels. The discovery of the fall in apparent viscosity in narrow tubes by Fahraeus and Lindquist [53], as well as the first measurements of blood viscosity in rotational viscometers by Brundage [19], were important milestones in the history of blood rheology. The dependence of blood viscosity on vessel radius, the phenomenon of plasma skimming and axial migration of cells, as well as their possible influence on this "Fahraeus-Lindquist-Effect" [5, 178] stood in the center of physiological interest. The phenomenon of red cell aggregation in many diseases and in circulatory shock, (sludged blood [71]) was of major clinical interest for many years. Bingham [11], one of the leading theoreticans of general rheology, emphasized another aspect of blood rheology, namely the unusually low viscosity of bulk blood when compared to other suspensions. This observation, however, was overlooked by physiologists in general, although MÜLLER [116] had also pointed to the unusually low resistance that red cells in high concentrations appeared to offer against passage through tubes of small diameter. BURTON (21), faced with the problem of zero flow in spite of a positive pressure head across certain vascular beds, studied the flow properties of red cells in protein-free ACD solution, and from this postulated that the blood itself was not capable of withstanding finite forces without flowing. This postulation was invalidated by later studies [108, 172] showing that the plasma proteins, mainly fibrinogen, exert a major influence on blood viscosity in prestatic flow and are responsable for the existence of elastic properties of blood in stasis (yield shear stress in vitro). Improved instrumentations in the early 1960's made possible measurements at extremely low flow. On the other hand, viscometry of blood was performed in very narrow tubes showing evidence of further reduction in apparent viscosity to values similar to plasma in vessels between 10 and 100 μm diameter. The physical properties of individual cells were also investigated using micropipettes [127, 128]. The principal attention concerned the characteristics of the red cell membrane and its deformability. The deformability of the whole red cell was analyzed by the filtration of erythrocytes through restricted pores, introducing a new concept for the understanding of a variety of hemolytic diseases and the removal of cells from the circulation [86, 87, 126, 162]. The theoretical basis of blood rheology was broadened by a number of workers from fields outside medicine, namely WHITMORE [181], DINTENFASS [40–45], MERRILL [107–113], and SCOTT-BLAIR [151, 152]. An important conceptional innovation was introduced by MASON [104]. Rather than studying the macroscopic flow behavior (bulk viscosity) of complex fluids under defined conditions, he described in detail the microscopic flow behavior of the particles or droplets constituting the dispersion. From this analysis he made predictions about the macroscopic flow behavior of the suspension or emulsion, thereby extending the now classical rheological studies by EINSTEIN [49], JEFFEREY [88] and TAYLOR [160, 161]. This extremely fruitful methodological approach was introduced into blood rheology by GOLDSMITH and MASON [60–67], and has led to a number of new concepts by these authors, as well as other groups, e.g. CHIEN and GREGERSEN (28–33, 72, 73), and WELLS and SCHMID-SCHÖNBEIN [142–145, 174]. It is obvious that in the field of hemo-rheology, microrheological considerations are of direct practical interest since they allow predictions about the flow of blood in the vast majority of all vessels.

Starting from the quantification and thence dynamic evaluation of red cell aggregation and its effect on the flow properties of blood, it soon became evident that the deformability of the erythrocyte is the major property responsible not only for the flow of blood in the capillaries and restricted pores, but also for the viscosity of bulk blood in large vessels at rapid flow. These microrheological studies therefore have revealed that the erythrocytes possess bipotential flow properties. Solely in response to alteration in the physical forces acting upon them, they can either be part of a three dimensional cell network with the rheological properties of a solid, or, when monodispersed, they can adopt the rheological properties of fluid drops and participate in flow. It becomes obvious that microrheological considerations are of direct significance to all cardiovascular function since they allow predictions of the flow of blood in the majority of all vessels. One might consider that the blood flow of the microcirculation has greater importance in the essential functions of cardiovascular systems than do the details of pressure and flow in the macro-circulation. The bipotential flow properties of red cells explain the reaction of the red cell to the physical forces acting upon it and its contributions to the three dimensional cell network present in blood both in stasis and flow. This report considers a number of studies which have been carried out to evaluate the hemorheological significance of these cellular variables, their dependence upon the properties and flow characteristics of the plasma and their influence upon blood flow and hemodynamics in general.

II. Principles of Microrheology (Newton's Law of Viscosity)

The flow of fluids is described by laws which relate motion of fluid elements to the frictional forces which occur as a result of the passing of one element over another. These elements are treated as layers or lamellae for the purposes of more precise mathematical integration. Their relative motion is caused by a force acting tangentially to the fluid layers (shear stress τ, dynes/cm^2) which is necessary to overcome the internal friction between these fluid lamellae: in other words viscosity of the fluid. As a result of this relative motion, the elements of the fluid are being deformed or sheared, a process that has often been compared (Fig. 1, top line) to the deformation of a stack of cards [113, 181]. The velocity of this deformation is equal to the velocity gradient between any two fluid lameallae: i.e. the velocity difference (du) divided by the distance (dr) of the two hypothetical fluid layers.

$$\dot{\gamma} = \frac{du}{dr} \left[\frac{cm/sec}{cm} = sec^{-1} \right] \tag{1}$$

This value is also called the rate of shear. The rate of shear (γ) and shear stress (τ) in simple fluids are proportional, the coefficient of viscosity being the proportionality factor (η)

$$\tau = \eta \cdot \dot{\gamma} \; [dyn/cm^2] \tag{2}$$

$$\eta = \frac{\tau}{\dot{\gamma}} \left[dyn \cdot sec/cm^2 \right]. \tag{3}$$

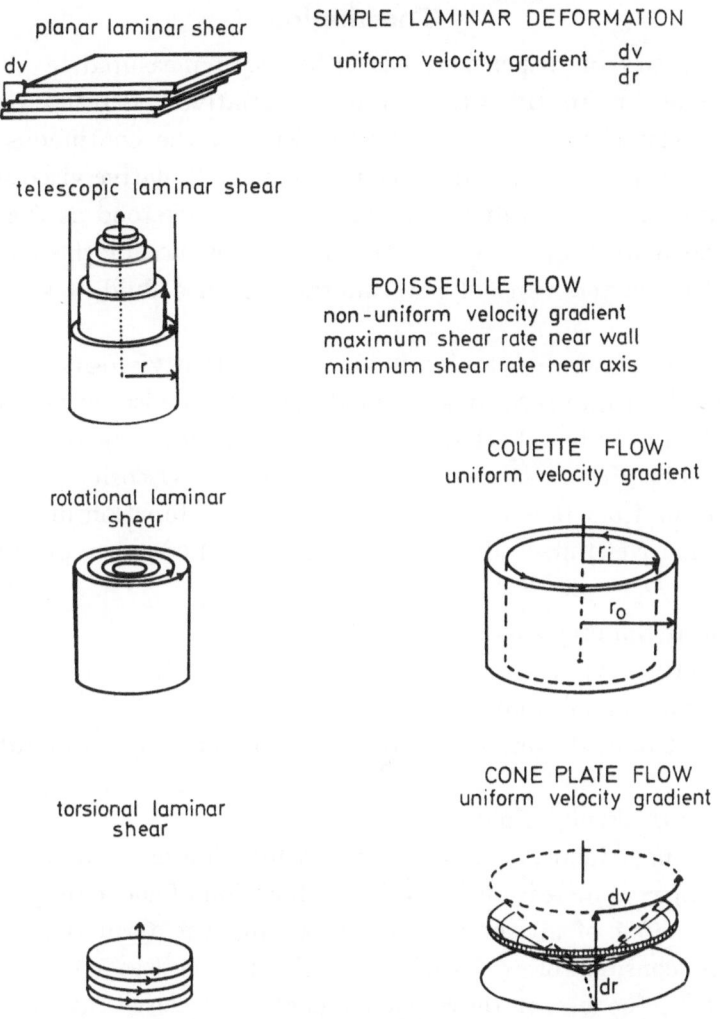

Fig. 1. Schematical representation of laminar shearing and deformation of fluid undergoing flow. For explanation see text

For simple fluids, the *coefficient* of viscosity—for any given temperature and pressure—is a constant, independent of the shear stresses prevailing. However, in the case of complex fluids, such as colloidal solutions, suspensions and emulsions, the *coefficient* of viscosity is not a constant but varies as a function of shear stress (or shear rate). Such fluids, which deviate from Newton's Law of Fluids (constant proportionality) are said to have "anomalous viscosity". Since there is no constant value of viscosity, the viscosity of such non-Newtonian fluids is described as the "apparent viscosity" for any given shear conditions. This can be visualized more clearly by plotting the apparent viscosity as a function of either shear rate or shear stress. A number of recent texts have appeared which can be consulted for more detailed rheological theory (4, 63, 113, 177, 181).

Terminology

The viscosity of suspensions has either been measured in absolute units (dyn · sec/cm² = poise (P) = 100 cP) or as relative viscosity (η_{rel}), i.e. the suspension viscosity divided by the viscosity of the continuous phase. Unfortunately, in physiology and medicine, the term "relative viscosity of blood" is very often used in a different fashion: it is understood as the viscosity of blood divided by that of water. This is consequence of the fact that most capillary viscometers can only be calibrated with a fluid of known viscosity (e.g. H_2O).

In the case of non-Newtonian fluids the apparent viscosity is calculated for each given shear rate according to equation 3. These viscosity data are plotted as a function of either shear rate or shear stress, covering the entire range of measurements. From viscosity profiles (apparent viscosity as a function of shear rate) or from flow curves (shear stress as a function of shear rate) the different non-Newtonian fluids have been classified on a phenomenological basis [131, 177].

1. [1] Shear thinning fluids:

a) Structural viscosity: decreasing viscosity with increasing shear rate. Synonyms: pseudoplasticity.

b) Thixotropy: decreasing viscosity as a function of shear rate *and* shear time.

2. Shear thickening fluids:

a) Dilatancy: increase in viscosity as a function of shear rate.

b) Rheopexy: increase in viscosity as function of shear time.

The meaning of all these terms, especially the term thixotropy, is still subject to considerable debate among rheologists (see e.g. 131, 177, 181). Beyond this, the use of these phenomenological classifications is of limited usefulness in the field of hemorheology. As will be demonstrated subsequently, the shear thinning of blood is not only a consequence of a breakdown of structure, but of both structural breakdown and structural rearrangement linked with phase separation (see p. 169). In measuring shear thickening suspensions, no clear distinction is possible between rheopectic and dilatant behavior [145].

Natural Flow in Cylindrical Tubes

In a rigid cylindrical tube of given length and radius, an incompressible fluid which is adherent to the wall be subjected to a pressure difference between the two ends. As—by definition—the fluid cannot resist a deforming force, and the outermost fluid layer is adjacent to the wall, the pressure difference induces a telescope-like sliding of the cylindrical fluid lamellae past

---

1 These and all following rheological terms are used according to the international nomenclature proposed by Reiner and Scott-Blair in 1968 (131).

each other: in other words, flow (2. row, Fig. 1). The resistance to this relative motion is overcome by shearing stresses acting parallel to the direction of flow.

Newton's law of fluids was first applied to the flow in cylindrical tubes by HAGEN [76] confirming POISEUILLE's experiments [123] which had established the correlation between flow rate and the 4th power of the radius. Simple Newtonian fluids show the parabolic velocity profile which results in a distribution of velocities of the lamellae across the tube and consequently a distribution of shear rates decreasing from a maximum value at the wall to zero at the axis. In tube flow, the measured quantities of driving pressure and flow rate depend for their interpretation on the knowledge of the relationship of the shearing forces producing this lamellar motion and the resulting velocity gradient or rate of shear. Since non-Newtonian fluids do not have a constant relationship between these two variables, the shear stress/shear rate relationship in such fluids therefore is not directly connected to the pressure-flow rate relationship.

Artificial Flow in Rotational Viscometers

For the reasons noted above, capillary viscometers are of limited usefulness in the study of complex, non-Newtonian fluids. If there is a distribution of shear rates across the tube, a distribution of apparent viscosity must likewise be assumed. This is likely to be more pronounced if there is a flow dependent separation of the continuous and dispersed phases (for example under the conditions of blood flow in narrow tubes, where an axial migration of erythrocytes occurs (see p. 207)). It is obvious that capillary viscometry can therefore yield only limited information about the causes of the non-linear relationship between shear stress and shear rate in blood.

It is, however, possible to subject a cylindrical fluid element not only to telescopic shearing in natural flow, but also to rotational and torsional shearing (Fig. 1, 3. and 4. row). Rotational shear is induced by subjecting the fluid to a stirring motion between the axis and the outer cylindrical surface of the fluid. Torsional shear is induced by a stirring between the upper and lower surface of the cylinder. All contemporary rotational viscometers utilize this principle. In such instruments the shearing of fluids introduces "viscometric flow" [34] which is easier to define than flow in cylindrical tubes. The co-axial cylinder viscometers (COUETTE), and the cone plate viscometers are the more common examples of such instruments. In COUETTE [38] viscometers, the fluid is placed between two parallel concentric cylinders and is subjected to rotational shear. Provided the distance between the two cylinders is small relative to the radii of the cylinders, a uniform and constant shear rate can be established in the fluid. This shear rate is proportional to the rotational speed of one of the cylinders. The shear stress is computed from the torque necessary to sustain this shearing. The apparent viscosity for each shear rate can be computed.

Especially for hemorheological measurements at shear rates near stasis (less than 1 sec⁻¹) Merrill with the cooperation of Dauwalter and Gillison [59] developed a viscometer in which artifacts due to the apparatus have been reduced to a minimum. A number of recent reviews have described and evaluated this instrument [113, 168].

Fluids can also be subjected to viscometric flow at a uniform shear rate between a stationary plate and cone rotating around its axis, i.e. a cone plate viscometer. In steady-state rotation, the velocities of the concentric fluid layers rise in a linear fashion.

As the layer adjacent to the plate is at rest and the layer adjacent to the rotating cone is moving with the velocity of the latter, a uniform velocity gradient is produced between the layers. If one considers any pair of points on cone and plate equidistant from the axis of the cone, the difference of their rotational speeds and the component of their distance from each other parallel to the axis of rotation will both be proportional to their distance from the axis. Therefore, the ratio of velocity difference to distance is the same for all pairs of points. Consequently, concentrical biconical fluid layers are in relative motion to each other. The resulting velocity gradient is therefore uniform in all parts of the measuring chamber and is directly proportional to the geometry of the viscometer and the rotational speed of the cone. As the measured value of torque is also directly proportional to the shear stress, the desired shear stress shear rate relationship can be derived directly from such measured quantities under a variety of velocity gradients.

Rheoscopy

The adaptation of a cone plate viscometer for direct microscopic study of red cells under shear allowed the correlation of shear stress measurements with the microscopic observation of blood in viscometric flow. A commercially available instrument (Wells-Brookfield microviscometer LVT, Brookfield Engineering Laboratories, Inc., Stoughton, Mass., U.S.A.) was modified for this purpose with regard to the cone and plate (Fig. 2). These were made transparent by machining them from methylmethacrylate (Plexiglas). The region near the tip of the cone is in very close proximity to the plate. Both the thin blood layer and low angular speed facilitate observation in this region, which, however, is subjected to the same rate of shear as the rest of the sample in the more peripheral regions of the measuring chamber. The viscometer, with transparent cone and plate were placed on the stage of an inverted microscope. The objectives were directed from below at the region of the cone apex. For a more detailed description of this instrument, see [142].

The microrheological studies to be reported subsequently were all based on microscopic observations of blood flow under specific rates of shear and correlated with the macroscopic (viscous) behavior of the blood in Couette and cone-plate viscometers. These techniques allow elucidation of the flow properties of erythrocytes and their influence upon the viscometric flow of blood.

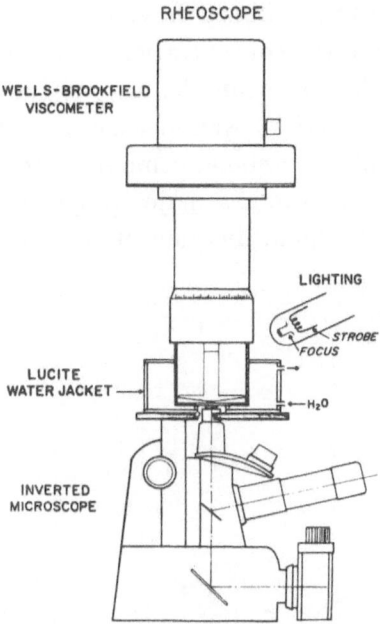

RHEOSCOPE

WELLS-BROOKFIELD
VISCOMETER

LIGHTING

STROBE
FOCUS

LUCITE
WATER JACKET

H_2O

INVERTED
MICROSCOPE

Fig. 2. Schematic representation of the "Rheoscope", rotational viscometer with transparent cone, plate and water-jacket are placed on top of an inverted microscope

III. Methods

The technical details of the viscometers used in these studies have been described previously in detail [59, 142, 171, 174]. The rheological data were plotted in two fashions: 1. as viscosity profiles and 2. as flow curves according to CASSON's [25] equation, as proposed by SCOTT-BLAIR (151). CASSON's equation was introduced to describe the flow behavior of certain printing inks in which a linear relationship exists between the square roots of shear rate and the shear stress.

$$\sqrt{\tau} = K_1 + K_0 \sqrt{\dot{\gamma}} \text{ where } K_1 \text{ and } K_0 \text{ are constants}$$

As first suggested by COKELET [36], it is possible to extrapolate the plotted flow curve back to zero shear rate, at which point the flow curve intercepts with the shear stress axis in a finite positive value.

$$\sqrt{\bar{\tau}} = K_1$$

COKELET [36] and BENIS [6, 7], among others have shown that this intercept represents the yield shear stress of blood

$$\tau_\gamma = K_1^2$$

Auxiliary Methods

In addition to the measurements of bulk viscosity of blood and the simultanous observation of the blood under shear (Rheoscopy), a variety of additional methods was used to study the flow behavior of red cell suspensions in which the physical properties and the concentrations of both cells and the suspending media had been altered. For example, the viscosity of concentrated cell suspensions (hematocrit 90–98%) was measured at shear rates between 1.15 and 460 sec^{-1} in a cone plate viscometer modified for this purpose [174]. The deformability of the erythrocyte is a major property for the passage of blood in the true capillaries. The quantification of this property is prerequisite for

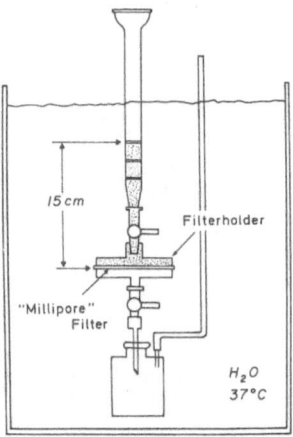

Fig. 3. Apparatus for measuring pore passage of erythrocytes modified after Murphy [118]. The filters are held in a micro-filterholder (millipore No XX30 025 00) which is connected to a burette and a recipient tube. The entire unit is kept at 37° C in a water-bath, the recipient tube is connected to open air

the analysis of red cell rheology. The ability of normal erythrocytes to pass through filter pores of 5–8 μm pore diameter is well established since the original experiments by Jandl [86], Prothero [126], and others. Erythrocytes rigidified by aldehyde or heat lose this ability [73, 162]. The passage time of erythrocyte suspensions through filters has therefore been taken as a measure of erythrocyte deformability (145). In the present experiments, a modification [145] of a filtration technique introduced by Murphy [118] was used. Dilute (2%) suspensions of erythrocytes in plasma were driven through Millipore-filters under standardized conditions (37° C, ΔP 15 ± 1 cm H$_2$O) and the flow time for 2 ml of this suspension was measured by a stop-watch. The following specifications for the filters used were given by the manufacturer (Millipore Corp., Bedford, Mass., U.S.A.): type SM, pore diameter 5.0 ± 0.3 μm, type SC, pore diameter 8.3 ± 0.3 μm and type NC, 14 ± 3 μm. The exact structure of these Millipore Filters has been subject to considerable debate. (see: Hemorheology, Proceedings of the I. International Congress on Hemorheology,

Reijkjavik, Iceland, 1967, Pergamon Press, New York 1969). Evidence is accumulating [125] that these filters are composed of irregularily spaced fibers and do not contain unbranched, cylindrical pores as claimed originally by the manufacturer. The polycarbonate sieves first used by GREGERSEN [73, 74] for this sort of study (Nuclepore, General Electric Corp., Pleasantown, Calif., U. S. A.) contain much more uniform pores and were therefore used in additional experiments in which 10 % suspensions of red cells in either 1 % isotonic albumin solution or plasma were studied in an apparatus as described by MURPHY [118].

In those experiments in which the influence of pO_2 and pCO_2 was tested, the blood samples were first equilibrated with the different gas mixtures. The same gas mixtures were then connected to the upper end of the burette of the filter apparatus. Thus both the atmospheres and the driving pressures could be altered.

Materials

Human venous blood was collected by gentle venipuncture from healthy blood donors. The following anticoagulants were used: 1. CPD solutions (67.5 ml of CPD solution to 450 ml of blood by routine blood bank procedures. The CPD solution had the following composition: 0.327 g of hydrated citric acid, 2.63 g of dextrose for 100 ml of solutions); 2. heparin, 100 USP-units per ml and 3. EDTA (1.5 mg/ml of Di-Na Ethylenediamine tetraacetate). If the tests were carried out within 4 h after venipuncture, no difference of the rheological behavior was observed in the samples using different anticoagulants. After 4 h, however, the filtrability of heparinized erythrocytes samples was markedly reduced, as was that of CPD blood after 24 h. For this reason, the use of heparin was discontinued and the test were carried out in EDTA or CPD blood samples within 8 h after withdrawal. In addition, blood from a number of hospitalized patients, anticoagulated with EDTA, was used (Myocardial infarction, surgical trauma, various hemolytic anemias)[2].

Preparations of Erythrocytes

Besides normal biconcave erythrocytes, swollen or crenated cells were used which were produced by washing the cells in 0.25 % albumin in a NaCl-phosphate buffer solution of the desired osmolarity (160–900 mOsmol/litre). After centrifugation, these pre-swollen or pre-crenated cells were mixed at hematocrits between 2 and 98 % with either plasma or 1 % albumin solutions of the

2 Most procedures were repeated on normal blood samples which had not been mixed with anticoagulants at all, to exclude effects of these agents. Viscometry and rheoscopy was carried out immediately after venipuncture and before the onset of coagulation. Packed cell viscosity and red cell filtrability were tested in defibrinated blood samples which were centrifuged at 0° C. The cells were then separated from the plasma and washed in 0.25 % isotonic albumin solution before coagulation occured. In all these cases, no significant differences in rheological behavior were detected when compared to those of EDTA anticoagulated blood samples.

same osmolarity. Concentrated suspensions of erythrocyte ghosts werde produced by freeze-thaw hemolysis of packed red cells [145] and then washed repeatedly as described by Goldsmith [62]. Rigid erythrocytes were produced by fixation of erythrocytes in 1.24 % isotonic glutaraldehyde solution for 24 h [145]. After repeated washings, the fixed cells, as well as the ghost preparations, were mixed with plasma or saline [146, 174].

Control of Hematocrit

Hematocrit controls in normal, swollen and crenated cells as well as in blood samples obtained from patients were carried out in microhematocrit tubes at 11,500 RPM for 2 min, the values were *not* corrected for trapped plasma. The volume fraction of rigid erythrocytes [145] see above) were determined by the [131]J-albumin-technique as described by Chien et al. [28, 32]. In suspensions of erythrocytes in highly concentrated dextran solutions ([144, 147] see below) the cells and the continuous medium were so similar in specific gravity, that centrifugal separation was impossible. Using precision syringes, the volume fractions of packed cells and continuous medium were fixed as precisely as possible and only the number of cells/mm³ was determined using a Coulter counter (Coulter Products, Inc., Hialeah, Fla., USA). The mean cellular hemoglobin concentrations of the red cells were determined by measuring total hemoglobin concentrations and hematocrits of the blood samples.

Suspending Media

Besides plasma, and homologous serum, a number of artificial suspending media were used for the present experiments. Dextran, with an average molecular weight of 40,000 (kindly supplied by Pharmacia, Piscataway, N. J., U. S. A.) was dissolved in concentrations between 3 and 35 % in distilled water. After adjusting the osmolarity to 300 m Osmols/litre by the addition of NaCl, the desired viscosity level was adjusted by the addition of further dextran or isotonic saline solution. Similarily, a 25 % albumin solution (American National Red Cross, Normal Serum Albumin [human]) was brought to isosmolarity and then used as continuous medium. The same albumin was used in 1 % concentration in isotonic saline in viscometric and in filtration experiments. In other experiments, the viscosity of the plasma was either reduced by adding an isotonic saline solution or was increased by adding a 10 % solution of Dextran 40,000 in isotonic saline.

The osmolarity of the suspending plasma or albumin solutions was increased by the addition of a Na-Cl-phosphate buffers. Hypotonic plasma was produced by diluting to the desired level with distilled water with subsequent reconstitution of the original protein concentration by ultrafiltration against vacuum using diaflo filter tubes. The total plasma protein concentration was determined photometrically using the biuret method. Plasma osmolarity was checked by freezing point depression (Advanced Instruments Osmometer). Advanced Instruments Corp., Newton Highlands, Mass., U.S.A.). For this the plasma was separated from the preswollen or preshrunken cells after osmotic equilibration. Changes in pH were produced by adding lactic acid and by equilibrating blood samples with high pCO_2. Lactacidosis was produced by adding lactic acid to the cell free plasma yielding a pH between 7.0–5.0. To the acidified plasma, red cells were added carefully and the pH of the suspension was then checked in a Beckman physiological gas analyzer (Typ 160). Hypercapnia was produced by equilibrating the blood sample with appropriate gas mixtures (10 % CO_2 in either N_2 or O_2) saturated with H_2O at $37°$ C. The measuring chambers of the viscometers and the filterapparatus were flushed constantly with the same gases during the actual rheological measurements. The blood pH and pCO_2 were again controlled in the above mentioned analyzer.

IV. Flow Properties of Erythrocytes in Plasma (Whole Blood)

Viscometric flow of whole blood in the transparent cone plate viscometer at rates of shear between 0 and 50 sec^{-1} is governed principally by reversible aggregation of erythrocytes (Fig. 4). Primary aggregation of red cells into rod-like rouleaux and secondary aggregation of rouleaux into branched, 3-dimensional aggregates are observed. More or less cell-free plasma gaps are seen between these larger aggregates. The size of these aggregates, and thus the plasma gaps, depends upon 1. the tendency to aggregation (see page 203), which is a function of cell shape and plasma composition; 2. the hematocrit and 3. the incident rate of shear. Upon each decrease of the shear rate, the size of the aggregate increases and vice versa [14].

At higher magnifications it became evident that even at a constant shear rate the larger networks continuously lost individual cells, as well as small aggregates, while simultaneously incorporating other cells and groups of cells. Lymphocytes and leucocytes do not appear to participate in this aggregation. Following a sine wave like variation of the velocity gradients [142] definite elastic behavior of these aggregates could be observed. During an increase of the velocity gradient, an elastic elongation of the rod-like rouleaux could be observed (Fig. 5). This elongation was a consequence of 1. a mutual sliding of the cells from a parallel into a slanting position, 2. a trapezoid deformation of the cells located at the branching point of two rouleaux and 3. an ellipsoid deformation of the cells within the extended rouleaux. The length thereby increased up to 3-fold. The slanting, as well as the ellipsoid deformation, eventually led to a pin-point connection between two cells. Upon further increase of the shear stress, the contact between any pair of cells within the extended rouleaux ruptured and the two parts of the rouleaux recoiled elastically. This recoil was the consequence of the resumption of the disc-like shape of the individual cells, of their relocation in parallel and often an S-shape bending of the remaining rouleaux. A similar recoil of the rouleaux was observed when the shear stress was reduced before rupture of the rouleaux occured. These elastic extensions, ruptures and compressions were seen only in those rouleaux

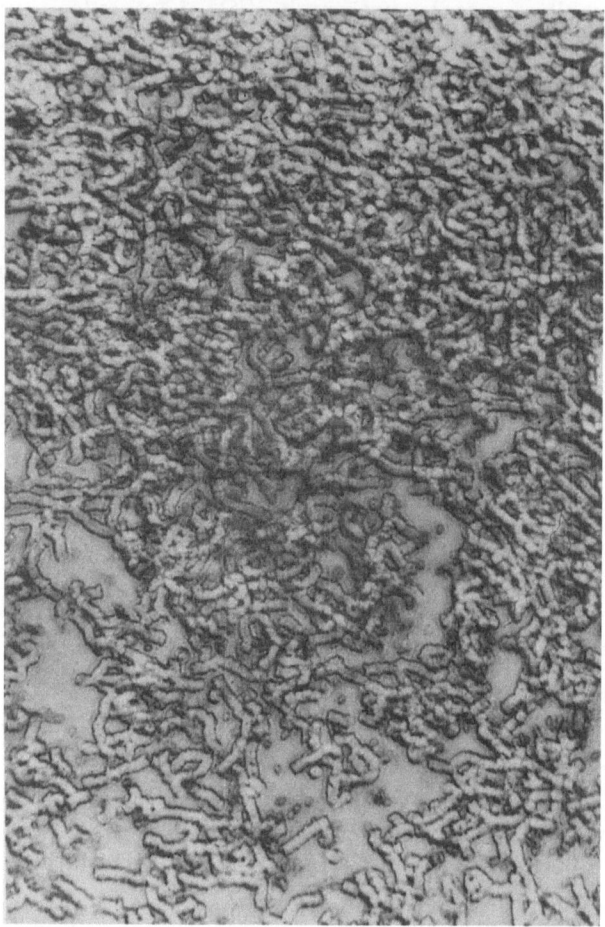

Fig. 4. Microscopic picture (strobe photomicrograph) of whole blood flowing at 2.3 sec⁻¹ in the transparent cone plate viscometer (Rheoscope). The majority of the red cells are aggregated into typical rouleaux, the rouleaux are aggregated into 3-dimensional networks

that were connecting two large aggregates and were bridging a plasma gap. In the majority of rouleaux which had multiple inter-connections within the network, no such elongation was observed: the whole network appeared to behave as one solid structure.

Aggregate Dispersion

Each increase in shear rate resulted in a partition of large aggregates into smaller ones, which was the consequence of the extension and rupture as

described above. Between shear rates of 5.8 and 46 sec⁻¹ each doubling of the
shear rate resulted in a reduction of aggregate diameter by about 50% and vice
versa. This relationship was valid only down to aggregate sizes of 15–30 μm in
diameter (Fig. 6) and was only observed immediately after a change of shear
rate had been applied. If prolonged viscometric flow at low shear rates was

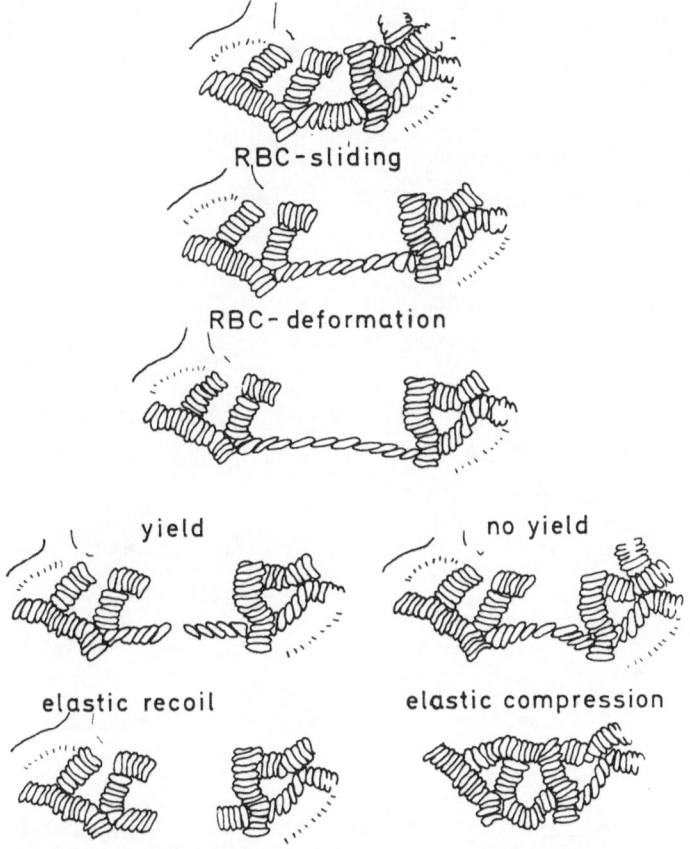

Fig. 5. Schematic drawing of the elastic behavior of rouleaux during increases of shear stresses. A mutual
gliding, deformation and slanting of cells within the rouleaux cause an elastic elongation

maintained, the aggregates tended to settle to the bottom of the measuring
chamber, where they participated less and less in flow and at the same time
increased in size. Once the aggregates were dispersed down to an average
diameter of about 15 μm, flow of individual cells, as well as of short spherical
and longitudinal aggregates (rouleaux) was observed. The latter were seen
to be much more resistant to further increases in shear. They rotated
as units in an irregular fashion around all axes. As in the case of larger
aggregates, reduction of the shear rate immediately (within 1–3 sec) resulted in the reformation of the larger networks. The heterogenous behavior
of the cells and short aggregates made it impossible to determine exactly

the shear rate at which the erythrocytes of a blood sample were entirely mono-dispersed. In order to obtain a quantitative measurement of the mechanical resistance of aggregates to dispersion, an arbitrary limit of 15 μm

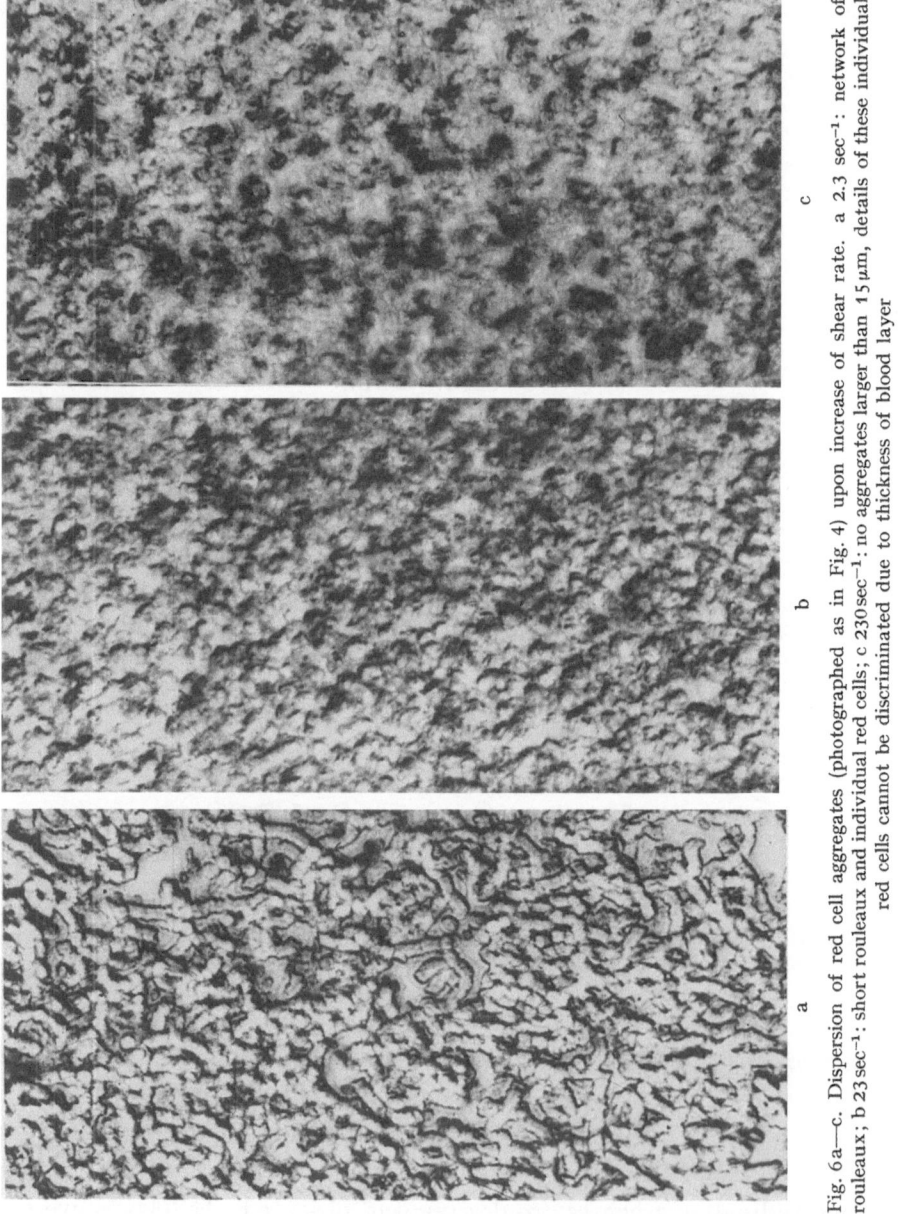

Fig. 6a—c. Dispersion of red cell aggregates (photographed as in Fig. 4) upon increase of shear rate. a 2.3 sec⁻¹: network of rouleaux; b 23 sec⁻¹: short rouleaux and individual red cells; c 230 sec⁻¹: no aggregates larger than 15 μm, details of these individual red cells cannot be discriminated due to thickness of blood layer

diameter was taken for a "dispersed aggregate" (141). By direct microscopic inspection as well as by stroboscopic photomicrography the picture of the flowing blood was observed and the shear rate determined at which no aggregates larger than 15 μm in diameter were seen to be present. This shear rate was termed the "maximum shear rate of aggregation" (γmax). Under the conditions of constant hematocrit and temperature, it is also possible to measure the shear stress necessary to disperse the aggregates in viscometric flow. For a 45 % blood sample they are about 2–4 dynes/cm² (69).

Each gradual reduction of the shear rate led to immediate reaggregation as described above. If, however, after aggregate dispersion the shearing was stopped abruptly, the individual cells remained monodispersed and re-aggregation was observed only after resumption of slow shearing. This indicated that aggregation is not due to an active attraction between cells but rather a consequence of cell collision, brought about by external forces (forces of flow or forces of gravity). The described red cell aggregation was observed without exception, in more than 500 different blood samples, both with or without anticoagulants. It must therefore be regarded as a physiological feature. The aggregates were dispersed more easily if the plasma was diluted by the addition of $^1/_3$ of the volume of 3.8 % Na citrate to blood which is customary in the determination of the sedimentation rate (modo WESTERGREN). This fact may explain why the physiological aggregation is so frequently overlooked [141]. The dependence of aggregation upon passive collision also explains why rouleaux formation is not seen in the normal blood smears. The very act of preparing a blood smear includes a rapid shearing of blood resulting in aggregate dispersion. If observed in a wet preparation of cells in their plasma, rouleaux formation and aggregation of rouleaux is seen invariably.

Pathologically Increased Aggregation

In almost all diseases in which the suspension stability of the blood is reduced, an intensification of the physiological aggregation was observed in viscometric flow [175]. In two groups of 72 patients suffering from acute myocardial infarctions and recovering from surgical trauma, all degrees of intensified aggregation were seen. These aggregates were dispersed only at shear rates above 230 sec⁻¹. At lower shear rates the aggregates were larger than in controls at any specific shear rate.

Although no basic or qualitative differences exist between these pathological and the physiological aggregates (both are shear dependent and reversible), a number of quantitative differences merit a separate discussion. In either case, rouleaux were the primary units of the aggregates. Under physiological conditions, they tended to be aligned in an end-to-side fashion while under pathological conditions a side-to-side alignment prevailed. This resulted in large cell clumps, large plasma gaps and a rapid sedimentation (Fig. 7). At high rates of shear, these aggregates were much more resistant to the dispersing forces of flow, the maximum shear rate of aggregation (γ_{max}) was elevated to 115 to 690 sec⁻¹ [146]. The shear stresses at this maximum shear rate of aggregation were 10 times higher than the normal controls (16–30 dynes/cm²).

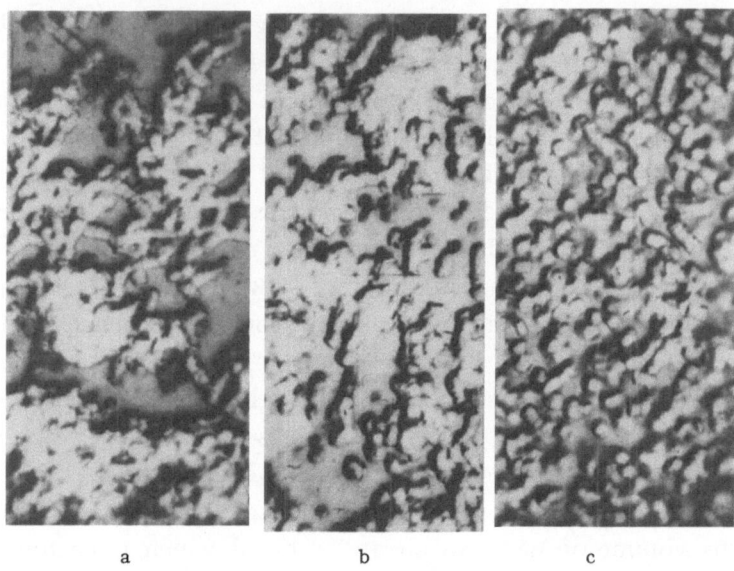

a   b   c

Fig. 7 a—c. Incomplete dispersion of pathological red cell aggregates in shear (photographed as in Fig. 4). a 2.3 sec⁻¹: large aggregates separated by clear plasma gaps; b 23 sec⁻¹: rouleaux and small aggregates; c 230 sec⁻¹: the majority of red cells is still aggregated into short rouleaux

Flow Behavior of Dispersed Erythrocytes at High Rates of Shear

At velocity gradients only slightly above the γ_{max}, rotation of individual cells was observed as described above. Details could only be observed in the immediate vicinity of the cone apex. In the more peripheral parts of the measuring chamber, in which the blood layer increased in thickness, no details could be observed. At velocity gradients above 230 sec⁻¹, strobe photomicrographs revealed that the cells had ceased to rotate. All cells appeared to be constantly aligned with their flat surface facing the transparent plate and in many instances, not a single cell was observed with its rim facing the plate. At a shear rate of 460 sec⁻¹ (Fig. 8) it became also obvious that the cells not only were constantly aligned parallel to the direction of flow but that they were being deformed from the customary biconcave shape during flow. As seen in Fig. 8, almost 100 % of the cells were aligned parallel to flow and were deformed into prolate ellipsoids, their long axis being parallel to the direction of flow.

A different fashion of deformation was observed when monodispersed erythrocytes in flow were observed in the more peripheral and thicker layers of the measuring chamber of the cone plate viscometer. Observations of individual cells in normal suspensions is impossible. However, following a method introduced by Goldsmith [62] 5 % intact erythrocytes were mixed with 50 % packed erythrocytes ghosts and 45 % plasma. The erythrocyte ghosts were transparent in transillumination, while they still mimicked the crowding effect of normal erythrocytes. In such "transparent blood" the consequences of mutual interaction of cells in flow became easily visible (Fig. 9).

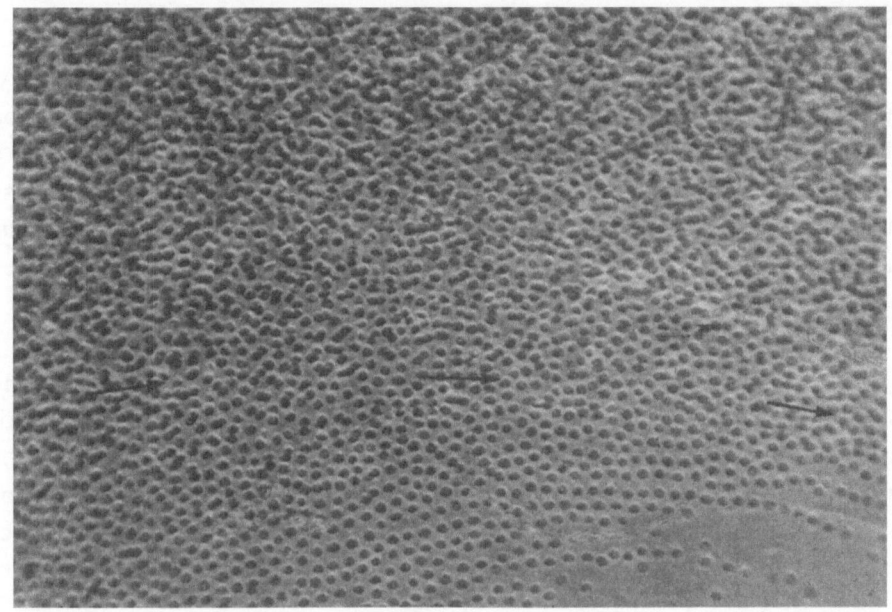

Fig. 8. Strobe photomicrograph of red cells flowing in transparent cone plate viscometer in the immediate vicinity of the cone apex. Shear rate 460 sec^{-1}. All cells face the bottom of the measuring chamber with a flat surface. The majority of the cells shows slightly ellipsoid shapes, the major axes of all ellipsoids are aligned parallel to the flow direction (arrow)

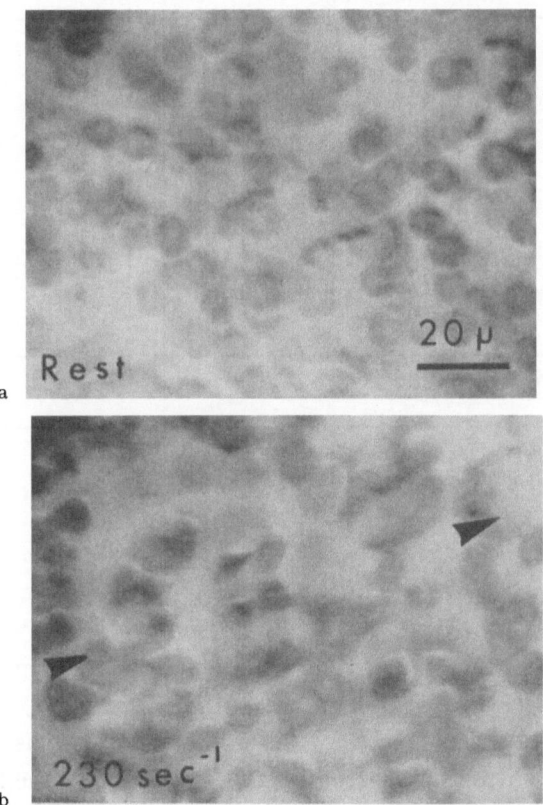

Fig. 9a and b. Strobe photomicrograph of red cells suspended in 45% plasma and 50% packed erythrocyte ghosts. a cells at rest are biconcave; b 230 sec^{-1}: the cells are deformed into irregular polyhedra

At low rates of shear, an occasional rouleaux formation of the intact erythrocytes was seen. At higher rates of shear the red cells were mono-dispersed, biconcave and rotated irregularly. However, at shear rates above 100 sec^{-1} the cells were no longer biconcave and were constantly changing their shapes in an irregular fashion. At 230 sec^{-1} none of the erythrocytes demonstrated biconcavity, a shape immediately resumed by all cells after lowering the velocity gradient.

Viscosity of Normal Blood

As reported by many authors [e.g. 25–33, 40–45, 107–113, 139–145, 171–175, 179] the viscosity of blood shows very pronounced shear rate dependence, although a certain variability exists between samples from different subjects. For any given temperature (37° C) and hematocrit value (40%) the apparent viscosity varies between 3.6 ± 0.33 cP at 230 sec^{-1} and 42.6 ± 11.4 cP at 0.1 sec^{-1} (Table 1, Fig. 10). The most pronounced increase in viscosity was

Table 1. *Apparent viscosity of bulk blood at hematocrits of 40% (n = 20) and 45% (n = 22) as a function of shear rate (mean and standard deviation)*

Shear rate (sec^{-1})	Hct 40% apparent viscosity (cP) ($\bar{x} \pm$ S.D.)	Hct 45% apparent viscosity (cP) ($\bar{x} \pm$ S.D.)
0.1 [a]	42.6 ± 11.4	67.7 ± 12.46
0.2 [a]	33.3 ± 12.0	48.6 ± 9.6
0.5 [a]	19.5 ± 6.1	25.3 ± 5.9
1.0 [a]	13.8 ± 3.3	17.3 ± 3.2
2.0 [a]	10.1 ± 2.2	12.7 ± 2.2
4.0 [a]	7.8 ± 1.4	9.6 ± 1.6
10.0 [a]	6.1 ± 0.92	7.1 ± 0.99
20.0 [a]	5.0 ± 0.66	5.8 ± 0.74
46.0 [b]	4.7 ± 0.43	5.6 ± 0.69
115.0 [b]	3.9 ± 0.39	4.6 ± 0.55
230.0 [b]	3.6 ± 0.33	4.1 ± 0.45
Extrapolated yield shear stress (dynes/cm²)	0.034 ± 0.016	0.046 ± 0.018

[a] GDM viscometer.
[b] Wells-Brookfield LVT viscometer.

seen in shear rates below 1 sec^{-1}. The shear rate dependence is even more pronounced at hematocrit 45% (Table 1), and increases with further increasing of hematocrit (see p. 200). In comparison to this physiological variability of blood viscosity and its dependence on hematocrit changes, the alterations of blood viscosity seen in diseases appear rather discreet [175].

As shown by Merrill et al. [108] this pronounced increase of blood viscosity at low rates of shear is only seen in suspensions of red cells in plasma, but not in serum. From this it has been concluded that the prestatic rise in apparent viscosity is the consequence of the interaction of erythrocyte and fibrinogen.

This assumption, repeating earlier ones by FAHRAEUS [52], was confirmed by studies in which the methods of rheoscopy and viscometry in a GDM viscometer were combined [142, 175]. Only erythrocyte suspensions which showed the primary and secondary aggregation described above exhibited the pronounced prestatic rise in viscosity at shear rates below 1 sec⁻¹ (Fig. 10). Suspensions of erythrocytes in normal serum showed occasional short rouleaux but no secondary aggregation, nor prestatic increase in viscosity. It can be

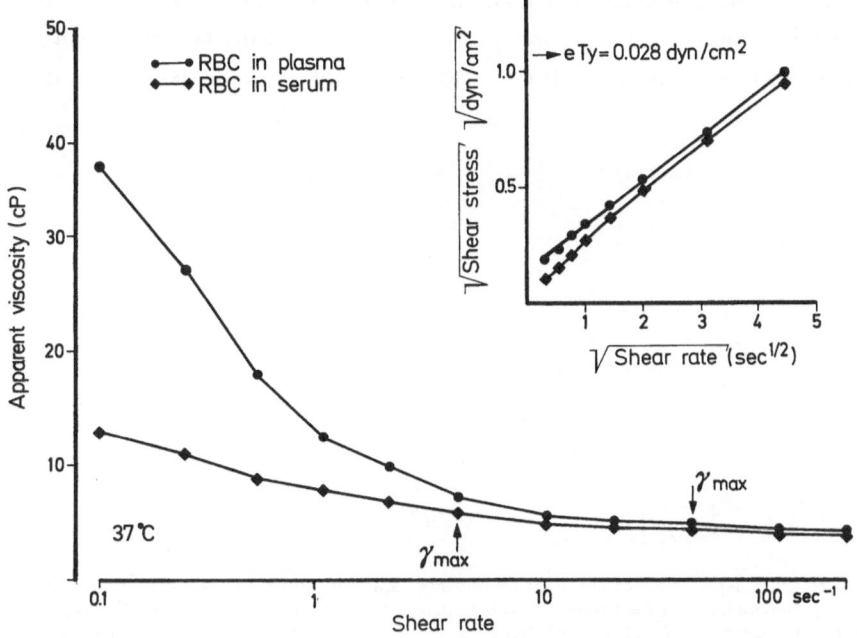

Fig. 10. Viscosity profile and CASSON flow curve of a normal blood sample (Hct 40%, 37° C) and of the suspension of the same red cells in serum

noted in parenthesis that under pathological conditions the red cells in serum may well exhibit both typical aggregation and pronounced prestatic increase in viscosity [175] identical to that observed in physiological plasma (see p. 203).

Extrapolated Yield-Shear-Stress of Whole Blood

Following a proposition by SCOTT-BLAIR [152], MERRILL et al. [36, 108–113] introduced the use of the CASSON equation ([24] see p. 155) for the clearer graphical representation of hemorheological data. As shown in Fig. 10 the rheological data of blood over a fairly wide range of shear rates lay on a straight line. CASSON had first demonstrated this linear relationship between the square-roots of shear rate and the square roots of shear stress for non-Newtonian printing inks. In these pigment—oil—suspensions, CASSON had supposed an aggregation of rod-like particles at low rates of shear and from this assumption he had deduced his flow equation. He was able to confirm this equation for dilute suspensions of pigment in highly viscous media and supposed an elongation of the aggregates with decreasing shear rate. Even though the direct observation of blood flowing at low rates of shear confirms the analogy of red cell rouleaux to Casson's rod-like particles, it is necessary to point out a number of rheological differences between these pigment oil and red cell plasma suspensions of rod-like particles; Firstly, red cells and red cell aggregates are suspended in *high* concentration in plasma, secondly these particles form continuous *networks* and thirdly, the viscosity of the continuous medium is *low*. Besides, very pronounced phase separation (separation of plasma and cells) is observed in blood, both in cone plate [141], and Couette viscometers [36]. Keeping these facts in mind, the Casson equation nevertheless offers itself as a

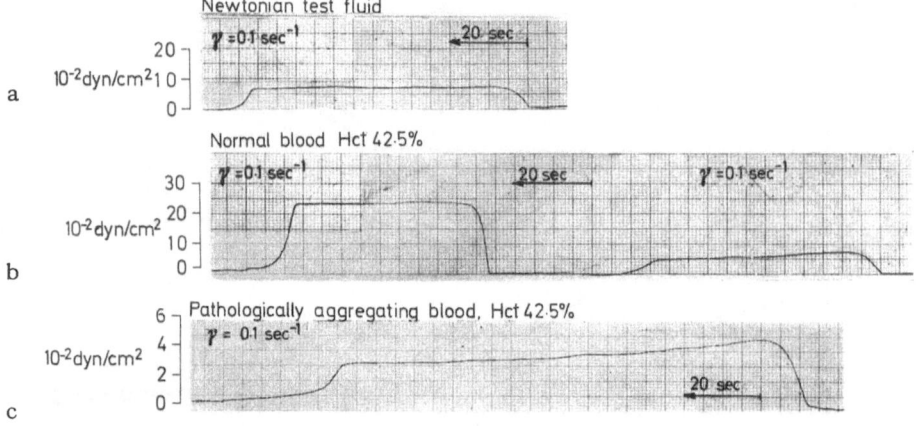

Fig. 11a—c. Time dependence of shear stress and consequently apparent viscosity of blood for a given shear rate. a Newtonian oil ($\dot{\gamma} = 1.0$ sec^{-1}, $\eta = 6.7$ cP, $\tau = 6.7 \times 10^{-2}$ dyn/cm^2. b Normal blood. At 1.0 sec^{-1} no time dependence (22.9 cP, 22.9×10^{-2} dyn/cm^2). At 0.1 sec^{-1}, the shear stress falls from $\tau = 7.1 \times 10^{-2}$ to 4.9×10^{-2} dyn/cm^2. c Pathologically aggregating blood. At 0.1 sec^{-1} the shear stress falls from 4.8×10^{-1} to 3.1×10^{-2} dyn/cm^2

convenient tool for describing shear stress/shear rate relationship for whole blood in viscometric flow. As first shown by Merrill's group [36], the intercept of the Casson flow curve with the shear stress axis represents the shear stress that whole blood is capable of withstanding without yielding or flowing. The square of this intercept therefore was taken as the numerical value of the yield-shear-stress of blood:

$$\tau y = K_1^2.$$

Using a number of different techniques it has been demonstrated that blood does indeed exhibit a yield-shear-stress [6, 7, 10, 26, 69, 110]. Although the numerical values vary considerably, they have generally been found to be in the same order of magnitude as that found by the extrapolation from the Casson flow curve. In non-aggregating blood samples, neither a linear Casson flow curve nor a positive intercept with the shear stress axis is found [142, 143]. More detailed analysis of Casson flow curves at shear rates below 1 sec^{-1} has shown, however, that in the wide bore COUETTE viscometers capable of measuring blood viscosity at these low rates of shear, a strong time dependence of the shear stress for any given shear rate, and thus a time dependent change in apparent viscosity, is observed (Fig. 11). It is very difficult to measure a constant shear stress value and the more pronounced the aggregation, the more pronounced the time dependence [175]. These effects, as well as the resulting deviation from ideal CASSONian flow behavior below 1 sec^{-1} has already been described in COKELET's initial communication [36] as well as in later reports by CHIEN et al. [29]. COKELET explained this deviation by a phase separation between aggregated red blood cells which occur at a faster rate than the establishment of a constant shear stress value at these low rotational speeds. He therefore extrapolated the shear stress values to the ones suspected at zero shear time i.e. immediately at the start of the viscometer rotation. The phase separation was called the "second heterophase effect" by MERRILL et al. [36, 109]. A similar phenomenon was observed in the rheoscope [141]; it leads to a falsely low shear stress and thereby low viscosity value, since shearing takes place predominantly in a low viscous plasma layer.

In spite of several attempts, it has not been possible to date to check this artifact in viscometric flow of blood [113]. Its effects upon blood viscometry are subject to differences in the geometric arrangement of the viscometers used: hydrodynamic forces of flow as well as gravity play significant roles. As pointed out above, phase separation in COUETTE and cone plate viscometers leads to shearing in plasma and thus falsely *low* viscosity values. In other viscometers, e.g. the cone in cone viscometers introduced by DINTENFASS [40] the slanted surfaces favor the sedimentation of aggregates [71] and thus lead to an increase in hematocrit within the measuring chamber thus resulting in falsely high apparent viscosity values [168, 175]. Another artefact that may be responsable for the unusually high viscosity values reported by DINTENFASS may be caused by surface artefacts (protein denaturation at the blood-air interface). This effect has been discussed by MERRILL [113] and WAYLAND [168]

For these reasons it is impossible at present to measure directly absolute shear stress values for blood at shear rates below 1 sec^{-1}; consequently the shape of the CASSON flow curve at shear rates below 1 sec^{-1} remains subject to debate. On the other hand, the very existance of a yield-shear-stress of blood can no longer be denied. The observations of the elastic properties of the red cell aggregates [63] in the rheoscope as well as the measurements of elastic response in bulk flow at low rates of shear [97, 167] substantiate the assumption of a yield-shear-stress. Furthermore, it can be observed that elements of blood, namely the described red cells aggregates can withstand shear stresses larger by several orders of magnitude (up to 3–30 dynes/cm^2 [146]), than the ones extrapolated for whole blood (40% Hct) in wide bore viscometers (0.02–0.08 dynes per cm^2 [175]).

As in the case of the γ_{max} (see p. 163), the extrapolated yield-shear-stress can only be taken as a true rheological value under closely defined conditions of specific viscometric flow and a specific composition of the blood [69]. In order to allow comparison between the viscometric flow behavior of different blood samples under identical conditions, it has been proposed [69, 175] to use the value of an "Extrapolated yield-shear-stress" (Casson), extrapolated for each viscometer from the linear part of the flow curve of blood for any given hematocrit. A similar proposition, although different in the evaluation of high shear viscosity (see below), has been made by Merrill [113]. This interpretation of data also takes into consideration comparative results by Charm [27] who extrapolated differing yield-shear-stresses of the same blood samples when measuring in different types of instruments (Cone-plate, Couette and capillary viscometer). This is not surprising in view of the fact that all forms of viscometric flow only represent an approximation of the model form of the linear shear deformation underlying Newton's law of viscosity.

Viscosity of Blood at High Rates of Shear

The apparent viscosity of whole blood did not reach an asymptotic minimal value after aggregate dispersion (Table 1). Even at shear rates in excess of 230 sec^{-1} the viscosity of blood continued to fall. Comparative viscosity measurements with a group of 22 blood samples in which the aggregates were dispersed at a shear rate of 52 ± 5 sec^{-1} showed without exception a lower apparent viscosity at 460 sec^{-1} ($93 \cdot 5 \pm 1.9\%$ of the values at 230 sec^{-1} [$p < 0.001$]).

Similar results for measurements in cone-plate viscometers were reported by Charm [27] whereas Merrill and Pelletier [112] postulated a transition from non-Newtonian viscosity of blood. More detailed analysis of Merrill's data shows that the transition into Newtonian flow behavior was deduced from observations made from *two* types of viscometers (Couette and capillary). Merrill conceded that secondary flow patterns occured in the GDM-Couette [59] viscometers above 315 sec^{-1}. It is impossible to exclude such phenomena at 315 sec^{-1}: it is entirely possible that at this shear rate, critical for the postulated transition, secondary flow behavior might have cancelled out the true fall in viscosity. The mode of plotting their data chosen by these authors (according to Casson's equation), minimizes the observed differences at high rates of shear. Since the question of a transition from Newtonian to non-Newtonian viscosity only applies to viscometric flow in large bore viscometers, it is of limited physiological relevance. Flow in small bore vessels in the microcirculation at high rates of shear is likely to be governed by the Fahraeus, Lindquist phenomenon [53], i.e. a very pronounced further decrease in apparent viscosity [3, 4, 58, 151] (see also Chapter IX, p. 206).

Red Cell Aggregation and Blood Flow

Using the technique of rheoscopy, it has been confirmed that red cell aggregation not only is a physiological feature of human blood but that it also increases apparent viscosity of blood very markedly at low rates of shear. These observations confirm those of many previous authors [52, 82, 172] but contradict KNISELY's opinion, who maintains [92] that erythrocytes physiologically repel each other and that aggregation is exclusively a pathological phenomenon. Using improved microphotographic technique, it has also been shown [146, 173] that erythrocyte aggregation is a physiological phenomenon in the venules of the human conjunctiva, vessels for which the lowest shear rates have been computed [146]. The novel technique of rheoscopy, however, has also shown that aggregation and disaggregation of erythrocytes cannot be the only cause for the anomalous viscosity of blood. Neither blood samples with very weak aggregation nor normal blood after aggregate dispersion have shown Newtonian flow behavior. In the range of high shear rates in which red cell suspensions were mono-dispersed they showed 1. further lowering of the apparent viscosity with increasing rates of shear, and 2. a pronounced cell deformation as observed in the rheoscope. The flow dependent cell deformations observed after mixing intact red cells with transparent red cell ghosts are very similar to those recorded by GOLDSMITH [65] and MONRO [115] in rapid tube flow (94 and 40 μm diameter respectively). All these observations gained additional weight in the light of CHIEN's discovery [30] that suspensions of non-deformable red cells were markedly higher in viscosity than suspensions of normal erythrocytes while at the same time they had lost their non-Newtonian flow behaviour. Since red cell aggregation at low rates of shear was always present in normal blood, the role of other possible red cell factors in blood rheology had to be studied in suspensions of red cells in artificial media. Red cells suspended in isotonic solutions of 1 % albumin or 3 % dextran 40000, osmotically crenated red cells in plasma and rigidified red cells in both plasma and saline did not aggregate and therefore offered themselves for such studies.

V. Flow Properties of Non-aggregating Red Cell Suspensions

Osmotic Crenation of Erythrocytes

The process of osmotic crenation of erythrocytes leads to very pronounced changes in the flow behaviour of their suspensions in plasma.

1. The tendency to aggregation is reduced or eventually abolished [143, 145]. This behavior is to be discussed more thoroughly on p. 198;

2. the viscosity of such suspensions is increased at all rates of shear, with emphasis on high ones;

3. the ability of the cells to pass through restricted filter pores is reduced.

At an osmolarity over 700 mOsmol/litre, 100% of the cells are transformed into crenated spheres, no aggregation is seen and the cells cannot be filtered through Millipore-Filters type SM (8 μm pore diameter) in spite of their reduced diameter. Fig. 12, compares the viscosity profile and the Casson flow curves of normal blood and of suspensions of crenated red cells in plasma at a hematocrit of 40%; the viscous effect of cell crenation is very obvious.

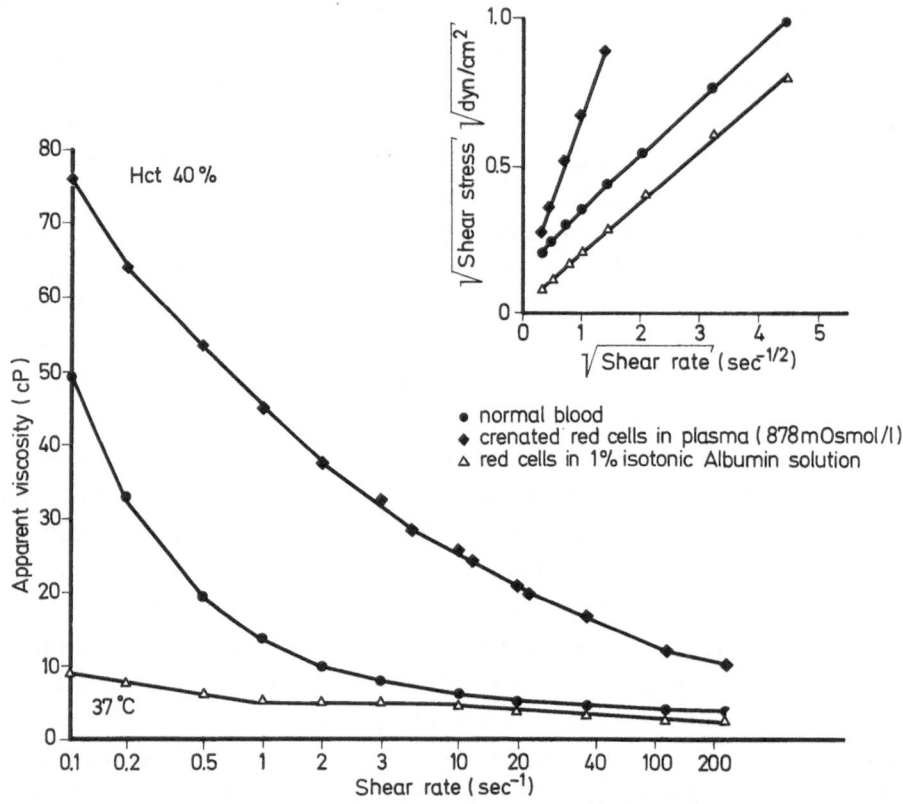

Fig. 12. Viscosity profile and Casson flow curve of normal blood, crenated red cells in plasma and normal red cells in 1% albumin solution. Hct 40%. Note increased viscosity of crenated cell suspensions and decreased viscosity of red cell suspension in albumin. A yield shear stress can be extrapolated only from the flow curve of normal blood

All these observations had been made separately by previous workers. Their correlation, however, was not established. This became possible by the simultaneous use of several methods, of which rheoscopy was most elucidating. It established beyond doubt that the observed increase in viscosity after osmotic crenation was *not* caused by aggregation. The microscopic evidence was substantiated by the changes of the Casson flow curves. In spite of a drastic increase in low shear viscosity, a yield shear stress could no longer be extrapolated, and the flow curve was bent.

Erythrocytes in Albumin Solutions

Erythrocytes suspended in an isotonic albumin solution (1%) were bicon-cave in shape, with no aggregation, neither at rest nor under shear. Their ability to pass narrow filter pores was practically identical to that of red cells in plasma (for details see p. 177). In spite of the absence of aggregation, the viscosity of a 40% red cell suspension was still shear rate dependent, falling from 9.0 cP at 0.1 sec^{-1} to 3.0 cP at 230 sec^{-1}. As shown in Fig. 12, the CASSON flow curve was bent, no yield shear stress could be extrapolated.

Normal and Crenated Erythrocytes in 3% Dextran Solutions

The very pronounced differences in the viscometric behavior of the two non-aggregating red cell suspension shown in Fig. 12, may at least in part be based on differences in the chemical composition and the viscosity of the suspending media. Such differences could be eliminated by using a 3% dextran solution which had approximately the viscosity of plasma (1.2 cP). In these solutions, the only difference was that of osmolarity. In spite of the absence of aggregation, suspensions of both normal and crenated cells were markedly non-Newtonian. However, the viscosity level of the crenated cell suspensions was always higher (for any given hematocrit and shear rate) than that of normal cell suspensions, the difference increasing with increasing hematocrit (Table 2).

Table 2. *Apparent viscosity of non-aggregated suspensions of biconcave and crenated erythrocytes in a 3% dextran solution (M. W. 40000) as a function of hematocrit and shear rate. Note the increased viscosity of crenated cell suspensions, especially at high hematocrits and high shear rates*

A. Normal erythrocytes in dextran 40000 1.2 cP, 300 mOsmol/litre

	Apparent viscosity (cP)		
	Hct 26%	Hct 51%	Hct 74%
0.1 sec^{-1}	3.86	27.2	124.0
1.0 sec^{-1}	3.39	18.1	52.5
20.0 sec^{-1}	3.10	6.6	15.5

B. Crenated erythrocytes in dextran 40000 1.2 cP, 600 mOsmol/litre

	Apparent viscosity (cP)		
	Hct 24%	Hct 50%	Hct 73%
0.1 sec^{-1}	5.73	47.8	296.0
1.0 sec^{-1}	4.78	27.7	138.6
20.0 sec^{-1}	3.82	16.2	80.5[a]

[a] at 4 sec^{-1}.

Rigidified Erythrocytes in Saline

The above mentioned findings indicated that in hemorheology the influence of cellular particles on the viscosity of the suspension does not agree with the accepted rules of suspension rheology. It is generally agreed that the viscosity of non-biological suspensions is primarily influenced by the volume fraction (hematocrit) of the suspended particles and not by their size [49, 164]. A very strong influence of cellular rheology became evident, when deformability of red cells (as measured by their ability to pass narrow pores) appeared to be such a critical variable in the viscosity of cell suspensions. Repeating the experiments of Chien et al. [30] the viscosity of non deformable erythrocytes after aldehyde fixation was measured. Such cells could not be passed through Millipore filters of 8.3 μm diameter. At a hematocrit of 30%, the viscosity of this cell suspension was independent of the shear rate. Such suspensions were essentially Newtonian fluids, a finding confirming Chien's et al. [30] original observations, as well as those of Seaman et al. [153]. When compared to the viscosity of normal erythrocytes in a 1% isotonic albumin solution, the viscosity of the latter was much lower, especially at high rates of shear [30]. After elevation of the hematocrit of fixed erythrocytes in saline, flow phenomena were observed which deviated from those described by Chien and which became even more pronounced when plasma was used as a suspending medium (vide infra).

Rigidified Erythrocytes in Plasma

Microscopic observation of rigid erythrocytes suspended in plasma showed no evidence of cell clumping or aggregation, neither at low nor at high rates of shear. During viscometry, however, marked changes in apparent viscosity, as a function of time, became evident (Fig. 13). In a 39% suspension, it was not possible to measure a constant shear stress value for any given rate of shear. Instead, a steady and reproducible increase in shear stress—and thus apparant viscosity—was seen. A value up to 100 times the value immediately after initial shearing was observed. After each mixing of the suspension, the shear stress value was identical to that observed initially. Shear thickening in such suspensions was also seen upon increasing the *shear rate* (Fig. 14). The details of this rheopectic and dilatant (see p. 152) flow behavior, not previously described in the hemorheological literature, have been published elsewhere [145]. Such phenomena are quite common in concentrated suspensions of non-biological particles [177]. They are considered to be the consequence of the interaction of the particles, resulting in the build-up of a structure by shear. While such structures may be easily broken down by gentle mixing, the observation in the rheoscope shows that they have nothing in common with the aggregation seen in normally deformable erythrocytes. Continuous rotation and frequent collisions were seen, but no sign of cell deformation nor cell alignment of any kind are noted.

Measurements reported by SEAMAN et al. [153] appear to be in conflict with those by CHIEN and the present ones. SEAMAN concluded that cell de-

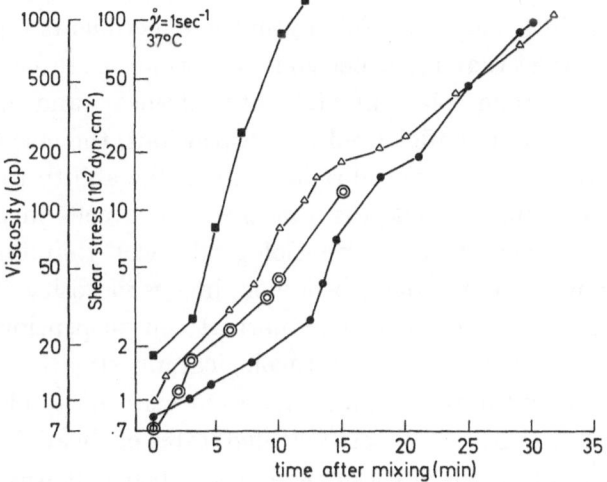

Fig. 13. Viscometry of rigidified red cells in plasma (Hct 39%). At a constant shear rate (0.1 sec⁻¹) a continuous increase of the measured shear stress value and thence the apparent viscosity are recorded. This phenomenon is fully reversible and reproducible in various samples (different samples are plotted)

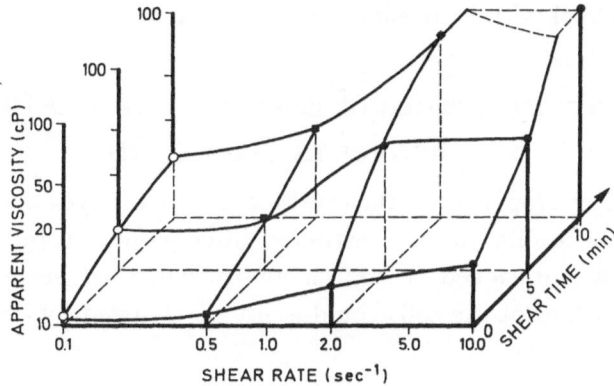

Fig. 14. Shear thickening as a function of shear time and shear rate in a suspension of rigidified erythrocytes in plasma. (After ref [145])

formability had a very limited influence on blood viscosity at normal hematocrit. It is very likely that the evidence presented for this conclusion is based upon incorrect interpretation of the hematocrit readings. Later experiments on hematocrit determination that SEAMAN performed in collaboration with CHIEN demonstrated the necessity of correction for trapped plasma between rigid cells [32]. Therefore, the hematocrit readings in [153] were erraneously too high.

The present findings, leading beyond the ones by Chien's group show that the normal deformability of the erythrocytes not only results in a very marked reduction of viscosity at high rates of shear, but appears to be *conditio sine qua non* for the flow of blood in tapering tubes. If normal blood also exhibited shear thickening, its flow in tapering tubes would interrupt itself, a phenomenon known as shear blockade [89]. From the experiments reported so far it must be concluded: 1. that in suspensions of normal erythrocytes in plasma, deformation of cells takes place at high rates of shear, and 2. whenever deformability is reduced in artificial red cell suspensions, the viscosity of these is markedly increased. After total cell rigidification, the ability to flow (fluidity) of concentrated cell suspensions was abolished. Such cells in plasma showed shear thickening rather than the physiological shear thinning. Osmotically crenated cells showed intermediate behavior: higher viscosity—or lesser shear thinning—at high rates of shear than normal cell suspensions. After these results, the prime interest of the hemorheological investigation presented here was no longer directed towards the high apparent viscosity of blood near stasis, but towards the low apparent viscosity at high rates of shear. Chien et al. [31] had shown that the high viscosity of rigid cell suspensions was comparable to the viscosity of latex particle suspensions: in other words it is a normal phenomenon. The mechanism, whereby the rheological properties of the normally deformable red cell allow a progressive reduction in apparent viscosity of aggregate free blood with increasing shear rate was to be investigated.

VI. Flow Properties of Erythrocytes as Affected by Changes in Osmolarity

The rheological properties of the erythrocytes themselves were investigated by studying their ability to pass restricted filter pores and by studying their flow behavior when packed in extremely concentrated suspensions. These rheological properties of the cells could easily be changed by osmotic swelling or by osmotic crenation.

Red Cell Filtration

The filtrability of erythrocytes, i.e. their ability to pass through pores that offer an inpermeable obstruction to non-biological particles of equal size is an observation of old standing. Erythrocyte filtrability in commercially available filters (Millipore, type SM 5.0 ± 1.2 μm average pore diameter, SC, 8.0 ± 1.4 μm and NC, 14 ± 3 μm average pore diameter) has been used since the original experiments by Prothero and Burton [126] and Jandl [86] as a measurement of cell deformability. In experiments carried out under conditions similar to those described by Murphy [118] (see p. 156), the erythrocyte filtrability showed a maximum value under isotonic or slightly hypotonic conditions (Fig. 15 and [105], [120]). In SC-filters, under a driving pressure of 15 ± 1 cm H_2O, the passage of a 2 % red cell suspension in plasma required 7.1 ± 0.2 sec ($n = 30$), 2 ml clear plasma required 5.0 ± 0.1 sec, the relative flow rate was 0.704 (Fig. 15). At osmolarities above 600 mOsmol and below 200 mOsmol/litre the relative flow rate fell to O. Similar results were seen

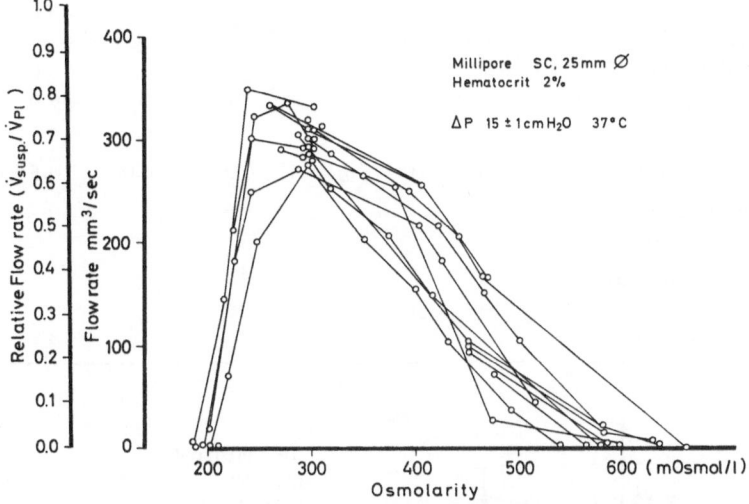

Fig. 15. Red cell filtration (2 % Hct $\varDelta$ P 15 cm H_2O, 37° C Millipore type SC, filterholder as described by Murphy [118]). Absolute (mm³/sec) and relative flow rate ($\dot{V}_{susp}/\dot{V}_{P_1}$) as a function of osmolarity

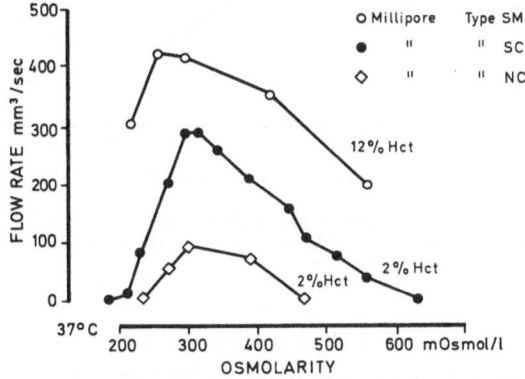

Fig. 16. Red cell filtration (as described in Fig. 15) using Millipore filter type SCg, SM and NC. A similar influence of osmolarity flow rate (mm³/sec) is seen in all filters

in SM and NC filters. In the latter the hematocrit was 12 %. A typical example is seen in Fig. 16. This shows that filtrability of human erythrocytes was reduced both by osmotic crenation and by osmotic swelling of the cells. Erythrocytes suspended in a 1 % isotonic albumin solution, at 2 % hematocrit filtered through type SC filters required 6.5 ± 0.4 sec for 2 ml at 15 ± 1 cm H_2O.

Table 3. *Influence of osmolarity on distribution of cell shape before and after filtration, relative flow rate (Millipore type SC) and percentage of cells sequestered in filter*

	Osmolarity mOsmol/litre				
	204	250	301	463	582
Cell shape before filtration					
a) biconcave	—	50	98.5	46.0	2.5
b) swollen	100	48	—	—	—
c) "crenated disc"	—	—	0.5	18.0	31.0
d) crenated sphere	—	2	1.0	36.0	66.5
Relative flow rate $(\dot{V}_{susp}/\dot{V}_{pl})$ SC filter	0	0.775 ± 0.04	0.737 ± 0.08	0.5 ± 0.1	0.05
Percentage of cells sequestered in filter	100	0	0	50	95
Cell shape after filtration					
a) biconcave	—	50	100	80	40
b) crenated spheres	—	—	—	10	32

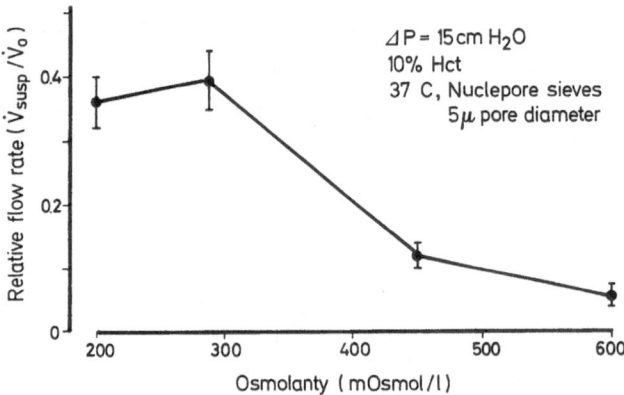

Fig. 17. Red cells filtration in Nuclepore sieves (5 μm pore diameters, 10% HCT, ΔP 15 cm H_2O, filter apparatus after Murphy). Influence of osmolarity on relative flow rate

It must be noted, however, that under the conditions of abnormal osmolarity, the measurement of the changes in filtration time do not exclusively reflect changes in average "deformability" of the cells. This is the consequence of the fact that the cells are affected by the changes in the osmotic environment in a non-uniform fashion. At 300 mOsmol/litre, all cells showed a biconcave shape, and 100% of the cells passed the filter. At 600 mOsmol/litre, where 100% of the cells were crenated spheres, no cells passed the filter. At intermediate osmolarities, biconcave discs, crenated discs and crenated spheres occured simultaneously, the percentage of the two latter abnormal forms increasing with increasing osmolarity. However, in the population of cells that had been filtered, there was always a greater percentage of "normal" cells than in the original population. Besides, the hematocrit of the filtrate decreased with increasing osmolarity. Therefore a preferential trapping of crenated cells in the meshes of the filters had to be assumed. It must therefore be concluded that the reduction in flow rate in SC and SM filters was at least in part caused by an occlusion of filter pores. Cell deformability therefore governs both the rate of pore passage of the individual cell and the number of pores available for perfusion. When using the polycarbonate sieves (Nuclepore, pore diameter 5 μm) for filtration of 10 % erythrocytes in plasma, a similar fall in relative flow rate is observed upon increase and decrease of osmolarity (Fig. 17). There is no evidence for sequestration of red cells in these filters; reduction of filter pores available for perfusion cannot be excluded, nevertheless. This ambiguous interpretation of filtrability data made desirable a second method to test deformability of red cells.

Packed Cell Viscometry

The very ability of red cells to be packed in concentrations of higher than 58 % is evidence that the cells must be easily deformable under the influence of gravitational forces (BURTON [22]). The fact, that packed cell suspension with

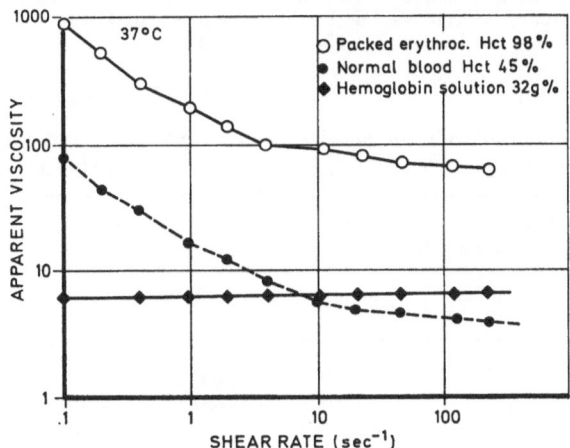

Fig. 18. Viscosity of packed erythrocytes (Hct 98 %), normal blood (Hct 45 %) and a hemoglobin solution (32 g- %) as a function of shear rate. Note that the viscosity of packed cells at rapid flow is lower than that of blood near stasis

hematocrits up to 98 % can be subjected to flow is a unique feature of blood rheology. All other known suspensions of particles lose their fluidity once the volume concentration exceeds about 60 % [45, 64, 164]. As shown by DINTEN-FASS [42] and WELLS et al., [174] the viscosity of these suspensions at high rates of shear can be taken as a measure of cell deformability. These studies have established a unique rheological adaptation of the erythrocytes under the flow conditions of extreme crowding by other cells.

The flow behavior of a 98 % red cell suspension is shown in Fig. 18. The viscosity is highly shear rate dependent, falling from about 700 cP at 0.1 sec^{-1} to 58 cP at 230 sec^{-1}. The original blood sample with 45 % hematocrit had a viscosity that was about $^1/_{10}$ of that of the packed cells at all shear rates.

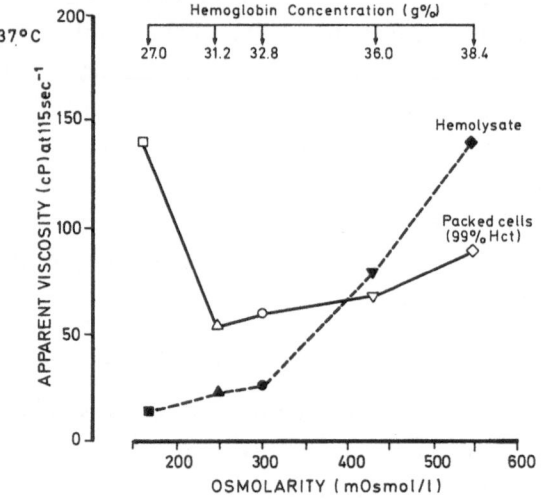

Fig. 19. Influence of osmolarity on apparent viscosity (115 sec^{-1}) of packed erythrocytes and their respective hemolysates. Note that after cell swelling, the viscosity of the packed erythrocytes is highest, while that of the hemolysate is lowest

Table 4. *Influence of osmolarity on hemoglobin concentration, and apparent viscosity as a function of shear rate of packed cells and their respective hemolysates*

Osmolarity (mOsmol/litre)		169	250	300	432	600
Hemoglobinconcentration (g- %/100 ml)		27.2	29.0	32.8	36.0	38.4
	Shear rate (sec^{-1})	apparent viscosity (cP)				
Packed cells	0.1	956.0	717.0	806.0	860.0	1 386.0
(99 % Hct)	1.0	253.0	148.0	186.0	215.0	425.0
	11.5	168.0	82.0	90.0	98.0	124.0
	115.0	142.8	58.0	65.2	66.6	86.0
Hemolysate	0.1	239.0	573.0	525.0	1 190.0	1 840.0
	1.0	81.0	176.0	181.0	368.0	610.0
	11.5	30.5	45.8	54.0	126.0	218.0
	115.0	14.4	31.6	29.2	78.0	148.0

A hemoglobin solution of equal concentration to packed cells was a Newtonian fluid of 5.9 cP. Osmotic alteration of the erythrocytes again had marked influence upon the packed cell viscosity (Table 4, Fig. 19). Slight osmotic swelling, between 290 and 250 mOsmol/litre, decreased packed cell viscosity, while further lowering of the osmolarity below 200 mOsmol led to very pronounced increases in viscosity. This biphasic viscous behavior was seen in spite of a continuous reduction of hemoglobin concentration. On the other hand, increases in the osmolarity regularly increased both hemoglobin concentration and packed cell viscosity. The described biphasic behavior was not seen when the red cell membranes were ruptured by freeze-thaw hemolysis. In these hemolysates the viscosity was a steady function of hemoglobin concentrations, rising from *hypo*-osmolarity through isosmolarity to *hyper*-osmolarity (Table 4). Although the volume fraction of the membrane fragments in the hemolysate is less than 3 % [22], these fragments had a very pronounced effect on the viscosity. Even traces of membranes increased the viscosity and led to non-Newtonian flow behavior (viscosity increasing with falling shear rate [37, 174].

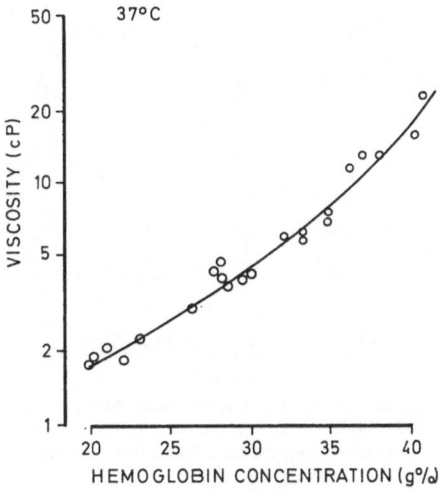

Fig. 20. Influence of the concentration of hemoglobin solutions upon their viscosity (shear rate independent). Note the strong increase of viscosity when hemoglobin concentration is increased above 32 g- %/100 ml

After careful removal of all cell fragments the remaining hemoglobin solution was a Newtonian fluid at all hemoglobin concentrations. The influence of hemoglobin concentration on hemoglobin viscosity is shown Fig. 20. The values correspond closely to the ones reported earlier by other investigators [37, 95]. The hemodynamic significance of the low hemoglobin viscosity will be discussed in Chapter VII, p. 191.

Microscopic Observations in Concentrated Red Cell Suspension

The optical density and hemogeneity of concentrated red cell suspensions precludes microscopic resolution of cell deformation. However, following GOLDSMITH's [62] suggestion, mixing 5 % intact erythrocytes in concentrated suspensions of red cell ghosts, allowed observation of the former. When seen at rest in the phase contrast microscope, all intact erythrocytes as well as the ghosts were deformed in a multitude of shapes. The normal biconcave red cell shape was resumed immediately after mixture with homologous plasma [174].

Using the bright field illumination of the rheoscope, only the intact cells were distinguishable among the packed suspensions of ghosts and cells. As in the phase contrast microscope, their shapes were quite irregular. Flow at the even lowest shear rates was accompanied by permanent deformation of the membrane. While the cells were moving in shear, certain surface designs migrated across the lower cell surfaces (facing the objective). Since this continuous cell deformation occured in very low shear rates, it could be studied very easily at magnifications up to 400 times. It thus became evident that during motion of the entire cell from right to left, the surface pattern of the membrane

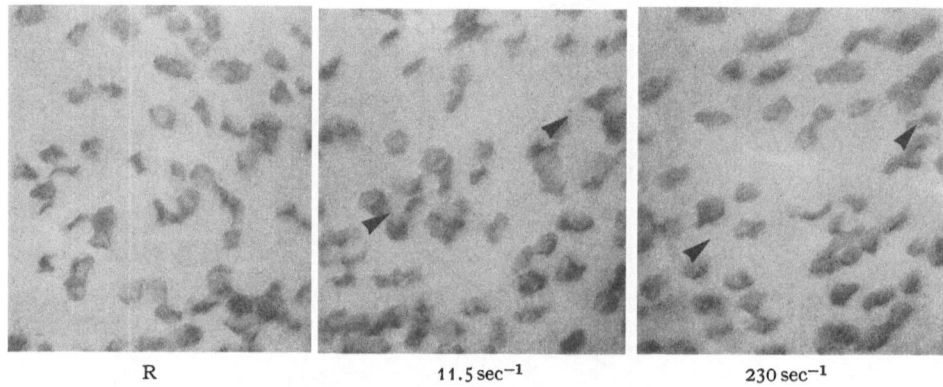

R   11.5 sec⁻¹   230 sec⁻¹

Fig. 21. Photomicrographs of erythrocytes in a packed suspension of ghosts. At rest all cells exhibit irregular shapes. At 11.5 sec⁻¹ the cells are being transformed constantly; at 230 sec⁻¹ the majority of the cells is aligned parallel to flow direction

described above moved from left to right. In other words, a rotation of the membrane around the cell content was seen. At very high rates of shear, the overall shape of the cells, still irregular, became more elongated and the major axes of the cells became permanently aligned parallel to the direction of flow. It is obvious that the continuous cell deformation taking place in packed cells at low rates of shear (which could be observed in slow motion) was similar if not indentical to the one photographed at high rates of shear in ghost-red cell plasma-suspensions (see p. 165 [66, 115]).

The physical properties of the red cells appear to allow a number of different fashions of flow and deformation. The ability of the cells to change their shape when passing through narrow channels, either in filter pores or capillaries in vivo (e.g. Brånemark [7], Bloch [13], Guest [75]) is likely to be based on the same characteristics that allow continuous membrane rotation. The specific shapes are likely to be a response of the deformable cell to the various shear forces acting upon it. In capillary flow, these forces act in all directions perpendicular to flow. In laminar viscometric or large tube flow, unidirectional shear stress acts parallel to the direction of the flow. The two opposite surfaces are subjected to viscous drag acting in opposite directions, to which the cell membrane responds by continuous rotation (see above). Hence the unusual phenomenon of packed cell fluidity appears to be the macrorheological correlate to the microrheological ability of the individual cell to undergo viscous deformation. This mechanism is a more plausible explanation of the phenomenon of packed cell fluidity than those presented previously. JACOBS [84, 85] had assumed that shear takes place in the remaining plasma layer. DINTENFASS [43] had presented the hypothesis that a shear dependent liquefaction and gelation of the membrane would allow the low "internal viscosity of the erythrocyte" to become hemodynamically significant. The observation of continuous membrane rotation gave a first hint that the deformable red cell reduces the viscous resistance of its suspensions by allowing the cell to participate in flow. This is a behavior typical for fluid drops and it is possible without a change in the physical properties of the membrane. The rotating membrane can be visualized to transmit shear stress from the outside to the inside of a cell, thereby inducing a system of laminar shear inside the cell. This would allow the closely packed cells to glide past each other with a minimum of frictional energy loss.

It can also be visualized that shear stress transmission and subsequent internal laminar flow might be inhibited by changes in the degree of osmotic inflation and deflation of the cell. The influence of osmolarity upon the fluidity of packed or individual cells can be explained as follows: Under the condition of *hypo*tonic swelling, the cell content becomes less viscous, which facilitates internal laminar flow. Under extreme inflation, when the membrane is under increasing extensional stress, its deformability is reduced and thus the rotation of the membrane under flow is inhibited. Thus shear stress transmission is in complete and a lesser internal circulation is established in spite of the fact, that the cell content is more liquid (see p. 203 [147]). Under the conditions of *hyper*-osmolarity, the cell fluidity can be inhibited due to increased hemoglobin viscosity. In spite of the fact that the membrane is now more deformable [127] the cell content is more viscous and less internal laminar flow is induced for any given shear stress [147, 174]. And indeed, when crenated red cells were dispersed in concentrated ghost suspensions they showed no deformation but rather tumbled in flow much like rigidified cells. It is obvious that this viscous adaptability of the cell, while first established for the flow of closely packed cells, is likely to be a most basic hemorheological mechanism. The viscous deformation and continuous membrane rotation would render to the cells properties of a fluid drop [39, 136, 137, 147]. Thus far, the proposed internal laminar shear—and therefore "red cell fluidity"—was still hypothetical. Substantial evidence favoring the concept of red cell fluidity was produced in experiments in which the laws of emulsion rheology were applied.

VII. Model Experiments on Red-Cell Fluidity

In case the membrane rotation observed under the conditions of tight erythrocyte packing in transparent ghosts should be acompanied by internal laminar flow of the cell contents, the erythrocyte would acquire the rheological properties of a fluid drop. If this is a general property of the erythrocyte, then red cell dispersions should behave rheologically like emulsions. The rheological behavior of fluid drops immersed in another fluid differs markedly from that of elastic particles. The rheology of emulsions has been defined by Taylor [160]; and his equations for this behavior have been confirmed recently [136, 136]. Taylor's [160] viscosity equation for emulsions is a modification of Einstein's [49] classical viscosity equation for suspensions. Einstein had assumed that the stream lines in the continuous phase were disturbed by the presence of solid particles. Consequently, the viscosity of the suspension was increased by a factor which depended upon the volume fraction of these particles (H), the viscosity of continuous phase (η_0) and the shape of the particles (K). Einstein's equation for dilute suspension is:

$$\eta_{\text{susp}} = \eta_0 + \eta_0 \cdot K \cdot H$$
$$= \eta_0 (1 + KH)$$

where the shape factor for spherical particles has the numerical value of 2.5. In concentrated suspensions the flow behavior cannot be described adequately by this equation due to effects of particle interactions. In the case of emulsions, Einstein's equation is modified by a factor T that depends upon the viscosities of both the dispersed phase (η_i) and the continuous phase (η_0).

$$\eta_{\text{emuls}} = \eta_0 (1 + KHT).$$

The term

$$T = \frac{\eta_i/\eta_0 + 0.4}{\eta_i/\eta_0 + 1}$$

has been called Taylor's factor [45]. Its numerical value varies between 0.4 when η_i/η_0 approaches 0 and 1.0 when η_i/η_0 approaches infinity. In the latter case the emulsion viscosity equals the suspension viscosity, since fluid drops with the very high internal viscosity behave like solid particles.

If one rewrites Taylor's equation to obtain the value of the relative viscosity (as described on p. 152)

$$\eta_{\text{rel}} = \frac{\eta_{\text{emuls}}}{\eta_0} = 1 + K \cdot H \cdot \frac{\eta_i/\eta_0 + 0.4}{\eta_i/\eta_0 + 1.0}$$

the following predictions can be made: If the red cell behaves like a fluid drop, the influence of increases in volume fraction upon relative viscosity should decrease with rising viscosity of the continuous phase. Experiments with artificial continuous phases of different viscosities support this hypothesis.

Failure of Red Cells to Increase Viscosity of Thick Dextran Solutions

Suspending red cells in a dextran solution of 60 cP (i.e. the viscosity of packed cells) resulted in a fluid of most unusual viscous behavior. The solution itself was an Newtonian fluid; after adding red cells (volume fraction under 50%, see p. 158) the suspension was non-Newtonian (Fig. 22).

The apparent viscosity fell from 482 cP at 0.1 sec^{-1} to 61.8 cP at 230 sec^{-1}. At this shear rate it was practically identical to that of the continuous phase. A 10% dispersion deviated from the viscosity of the continuous phase—only at shear rates below 1 sec^{-1}, the viscosity of a 25% suspension fell from 125 to 16.4 cP with increasing shear rate. Even in the 75% dispersion the viscosity

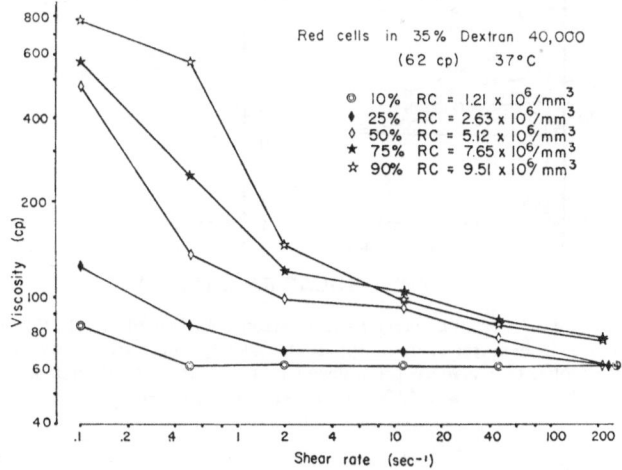

Fig. 22. Apparent viscosity of dispersions of red cells in a Dextran solution (MW 40000). Influence of rising cell numbers/mm³ on viscosity at various shear rates. Note the pronounced shear thinning of all dispersions

at low shear rates was 8 times that of the continuous phase but fell with increasing shear rate very markedly. At 230 sec^{-1} the viscosity was only slightly over 80 cP or 1.3 times that of the cell-free medium. In other words: at low rates of shear the addition of erythrocytes did cause the expected rise in viscosity. With increasing rates of shear, however, a shear thinning was produced so that at 230 sec^{-1} the erythrocytes no longer excerted viscous effects in the flow of the dextran solution. A similar finding was reported by DINTENFASS [45]; this author did not refer to the very pronounced shear dependence of the viscosity of these dispersions.

This unusual shear thinning was also found when using less viscous dextran solutions. Fig. 23 shows the influence of the red cell volume fraction upon the apparent viscosity of dispersions with a continuous phase viscosity of 1.2, 6.7, 12.0 and 45.0 cP respectively (all with the cells from one subject). As predicted by TAYLOR's equation, the influence of rising hematocrit upon dispersion vis-

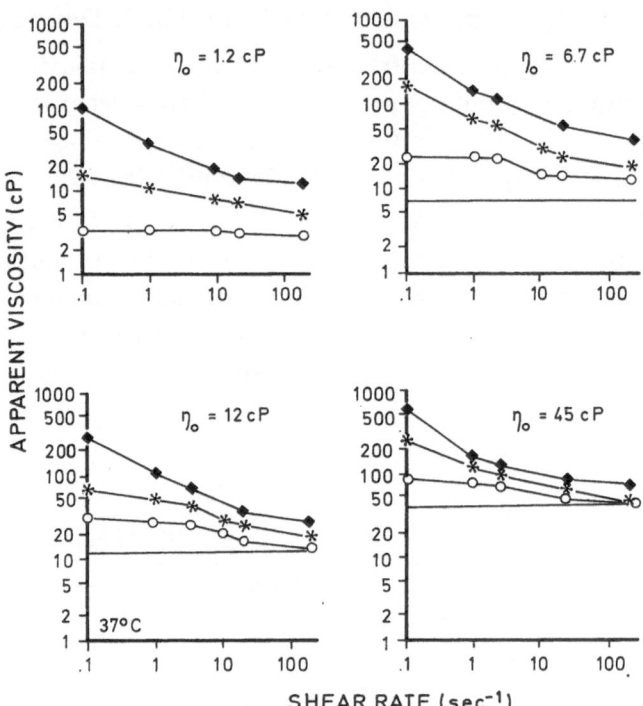

Fig. 23. Apparent viscosity of red cell dispersions in dextran solutions of 1.2 cP, 6.7 cP, 12 cP and 45 cP. The influence of increases in hematocrit upon apparent viscosity falls with rising viscosity of continuous medium. This effect is consistently more pronounced at high shear rated than at low ones. ○ = 25 % erythrocytes, ✻ = 50 % erythrocytes, ◆ = 75 % erythrocytes

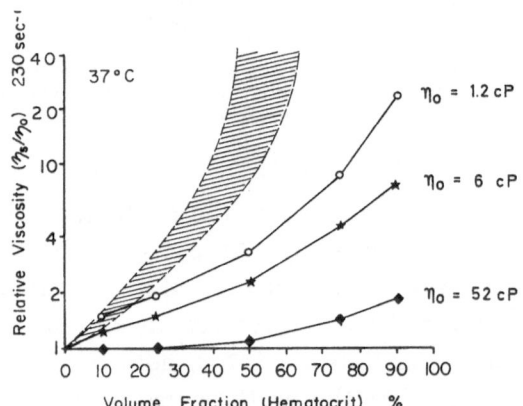

Fig. 24. Influence of hematocrit upon relative viscosity ($\gamma = 230$ sec^{-1}) of red cells suspended in Dextran solution of 1.2 cP, 6.0 cP and 52 cP. The shaded area represents the values for rigid particles as compiled by Thomas [164] from the work of 16 different authors. Note that at volume fractions above 40 % the relative viscosity of rigid particle suspensions rises exponentially. In comparison to this, the relative viscosity of red cell dispersions rises only slightly with increasing hematocrit

cosity fell with increasing continuous phase viscosity. In spite of the absence of red cell aggregation it was noted that all these red cell dispersions were highly non-Newtonian, and that only the less concentrated dispersions appeared to have reached an asymptotic viscosity value at 230 sec^{-1}, the highest shear rate investigated in the present series. At these high shear rates, the predictions from TAYLOR's equation were verified.

This is also demonstrated in Fig. 24, showing a similar experiment again using the erythrocytes of a single subject and dextran solutions of 1.2, 6.0 and 52 cP. The apparent viscosity at 230 sec^{-1} was measured and the relative viscosity (see p. 152) was plotted as a function of hematocrit. The predictions from TAYLOR's equation are again confirmed, the influence of the volume fraction of erythrocyte upon relative viscosity fell with rising continuous phase viscosity. In the same picture, the influence of solid particles upon relative viscosity is plotted for comparison. The shaded range includes values from the work of 16 authors that have been compiled recently by THOMAS [164].

Erythrocyte Deformation in Thick Dextran Solutions

When observing the flowing red cells in an isotonic dextran solution of more than 50 cP in the transparent viscometer, the cells both at rest and at low shear rates are seen to be monodispersed, exhibiting the familiar biconcave shape. At a shear rate of 1.15 sec^{-1} to 2.3 sec^{-1}, the cells remained biconcave, tumbling while moving in the viscometer. After each increase of the shear rate above 4.6 sec^{-1}, the flow appearance of the cells changed as follows: 1. the individual cells lost their biconcave form, were elongated into flat prolate ellipsoids and their longitudinal axis aligned parallel to the direction of flow; 2. tumbling of the cells was completely eliminated; the cells remained aligned parallel to the flow direction at all times, and 3. all cells were deformed and aligned in a similar fashion, resembling the appearance of a school of fish. Each further increase in shear rate intensified the described appearance, the cells becoming more and more elongated with each increment in shear rate. Immediately after a sudden stoppage of the flow, the cells resumed their familiar resting biconcave shape, while each stepwise reduction of shear rate resulted in shorter ellipsoids (Fig. 25). The occurence of this pronounced red cell deformation at very low rates of shear made it possible to study this process using higher magnifications (400×). The speed of motion was low enough, and the magnification high enough to recognize the following details. At 1.15 and 2.3 sec^{-1}, when the cells were still largely in their biconcave shape, the indentation disappeared and reappeared occasionally. At 4.6 sec^{-1}, the cells began to assume an elongated shape. The indentations were still present in these short ellipsoids, again disappearing upon slight stretching and reappearing after shortening of the major axis of the cell. At this magnification and at higher shear rates it became evident that the cells migrated away from the stationary plate towards the rotating cone. At further increase in shear rate, when the cells were fully elongated and transformed into ellipsoids, the membrane was seen in a rotating motion around the cell contents. This pattern of motion could best be observed when filtered India ink particles were added to the red cell dextran mixture. Carbon particles of 1–2 μm were seen attached loosely to the red cells and could be observed orbiting with the membrane for 1–3 rotations before becoming separated again. Similarly, larger carbon particles could be seen attached laterally to an erythrocyte and could also be observed in an orbiting rotating motion while the whole cell itself moved as an ellipsoid and remained aligned. At these high magnifications it could also be seen that the cells very seldomly collided. Collision usually occured when a more rapidly moving red

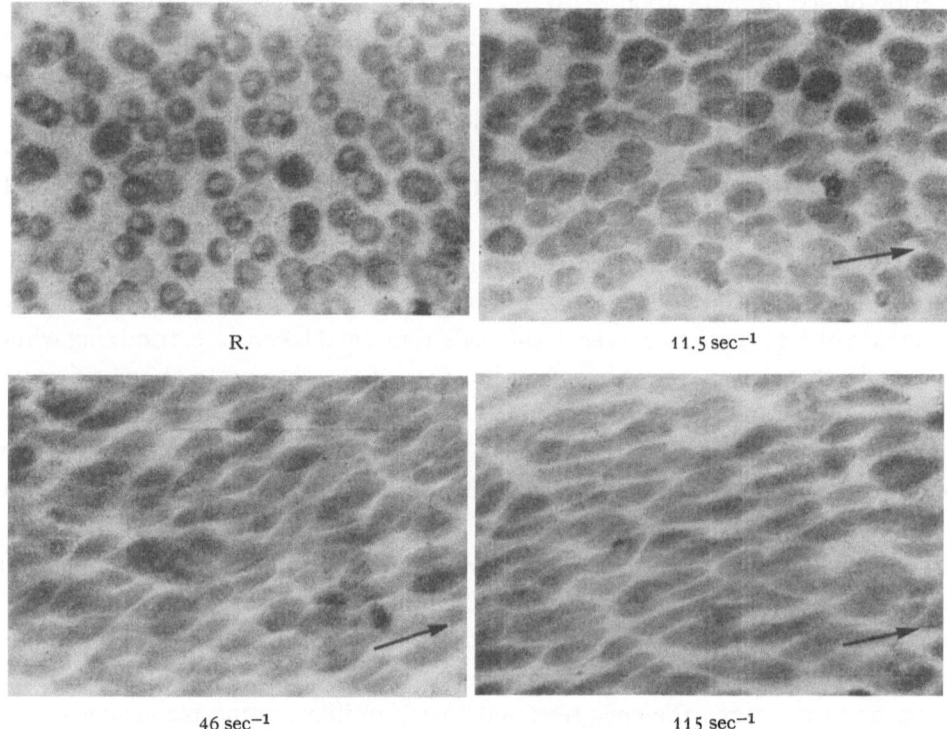

R.   11.5 sec⁻¹

46 sec⁻¹   115 sec⁻¹

Fig. 25. Strobe photomicrograph of red cells suspended in a dextran solution of 50 cP. Cells at rest are biconcave. With each increase in shear rate, the cells become progressively deformed into prolate ellipsoids, their major axis aligned parallel to the flow direction (= arrow)

cell touched a slower one. A very peculiar behavior resulted: the faster cell, upon touching the other, suddenly became markedly accelerated and seemed to leap shead of the slower cell. The slower cell in turn was only slightly slowed down by this process. No other interaction, temporary attachment or motion across planes of shear was observed upon cell collision.

Observations of the red cell suspended in dextran solutions of 30, 6 and 1.2 cP showed a similar behavior. At low shear rates the cells were always monodispersed. They became elongated into ellipsoids with increasing shear rate and were aligned parallel to flow. The most conspicuous finding, again, was the total absence of cell tumbling ("school of fish flow"). When a 1.2 cP dextran solution was used, cell deformation was only seen at shear rates in excess of 460 sec⁻¹. The rotational speed at these shear rates was too high to allow direct and detailed observation in the "Rheoscope". Strobe microphotography, however, again revealed deformation and alignment of the cells as described above. The ellipsoid deformation seen in red cells plasma at a shear rate of 460 sec⁻¹ (see p. 165, Fig. 8) can now be understood as the same fluid drop like deformation.

Effect of Red Cell Rigidification

Suspensions of totally rigidified cells resulted in markedly different viscous behavior. As seen in Fig. 26 and Table 5, the viscosity very markedly rose as

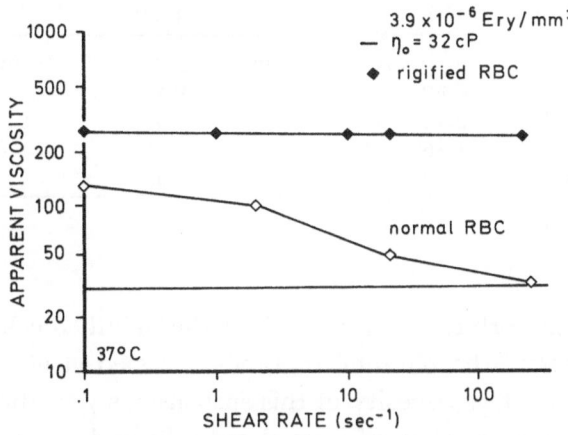

Fig. 26. Viscosity profile of normal and rigid erythrocytes suspended in 32 cP dextran solution (3.9 × 10⁶ cells/mm³). Note that due to red cell fluidity, the viscous effects of the normal cells upon dextran viscosity disappears at high rates of shear

a function of the hematocrit. At cell numbers in excess of 5×10^6, such suspensions cease to flow. Immersed in a dextran solution (30 cP) containing 3.8×10^6 rigid red cells/mm³, viscosity was 270 cP at all rates of shear (Fig. 26). In contrast to this, normal red cells in the same dextran solution had a viscosity of 151 cP at 0.1 sec⁻¹ with progressive shear thinning at higher shear rates, until, above 115 sec⁻¹, the viscosity was identical to that of cell-free Dextran. Analoguous behavior was seen at lower concentrations (Table 5). While the viscosity of rigid cell suspensions was independent of shear rate, that of normally deformable cells fell progressively with increasing rates of shear and approached the viscosity of the continuous phase at 230 sec⁻¹. Microscopic observation of rigid cells in dextran revealed their familiar biconcave shape and they were seen in continuous tumbling, orbiting irregularly with frequent collisions. Suspending of osmotically crenated cells in hypertonic (600 mOsmol/ litre) dextran solutions between 1.2 and 50 cP also resulted in a different flow behavior. At low rates of shear, the cells remained undeformed, i.e. the crenated

Table 5. *Relative viscosity of rigid and normal red cells suspended in a dextran solution of 28 cP. Influence of cell number and shear rate*

γ (sec^{-1})	Normal red cells ($\eta_0 = 28$ cP)	Rigidified cells ($\eta_0 = 28$ cP)
	7.08×10^5 RBC/mm^3	7.05×10^5 RBC/mm^3
0.1	1.26	1.23
1.0	1.18	1.53
23.0	1.16	1.72
230.0	1.10	1.78
	2.59×10^6 RBC/mm^3	2.68×10^6 RBC/mm^3
0.1	2.17	3.52
1.0	1.82	3.66
23.0	1.36	3.80
230.0	1.16	3.87[a]
	3.59×10^6 RBC/mm^3	3.79×10^6 RBC/mm^3
0.1	4.82	9.15
1.0	2.73	9.65
23.0	1.36	9.51
230.0	1.12	9.68[b]

[a] 115 sec^{-1}.   [b] 46 sec^{-1}.

spheres tumbled and orbited as described for the rigidified cells. Only at shear rates above 100 sec^{-1}, the photomicrographs revealed stretching of the cells into ellipsoids [147]. The viscosity of this suspensions was always higher than that of comparable suspensions of normal red cells in dextran solutions [147].

Rheology of Red Cells in Concentrated Albumin Solution

To assure that the observed effects were not caused by a specific Dextran effect, these experiments were repeated with a 25 % albumin solution as continuous medium. The same cell deformation as in dextran of equal viscosity was observed. Table 6 summarizes the influence of hematocrit on the relative viscosity of red cell albumin dispersions, the values are comparable to those obtained in Dextran solutions of 6.1 cP.

Table 6. *Relative viscosity of red cells suspended in albumin and dextran solutions. Influence of cell number and shear rate*

	η_0 centipoise	Hct %	RBC/mm^3 ($\times 10^6$)	η_{rel} at 2.3 sec^{-1}	η_{rel} at 230 sec^{-1}
Dextran solution	6.1 ± 0.7	25	2.35 ± 0.37	2.52 ± 0.94	1.91 ± 0.3
	($n = 9$)	50	4.27 ± 0.60	6.57 ± 1.10	2.36 ± 1.1
		75	6.48 ± 0.76	14.66 ± 2.30	4.55 ± 0.8
Albumin solution	5.1 ± 0.1	25	2.88 ± 0.35	3.98 ± 0.50	1.84 ± 0.2
	($n = 3$)	50	4.57 ± 0.88	9.43 ± 0.95	3.52 ± 1.1
		75	6.66 ± 0.20	15.75 ± 2.00	5.92 ± 1.6

Aspects of Erythrocyte Micromechanics

These studies have shown that the erythrocyte is indeed capable of assuming physical properties similar to those established for fluid drops. The present viscometric data are in accordance with those reported by DINTENFASS [45]. GOLDSMITH has described the ellipsoid deformation of erythrocytes in highly viscous dextran under the conditions of Poisseuille-flow [67]. The following discussions of the micromechanics of the erythrocytes can therefore be based on observations of actual erythrocyte fluidity under a variety of conditions. This fluidity is the consequence of a) the fluidity of the cell content, and b) the flexibility of the cell membrane, and c) the shape of the erythrocytes, i.e. the fact that the surface area to volume ratio is not minimized in a cell with the shape of a biconcave disc.

FUNG [54] has shown in a theoretical treatise that these three factors are the prerequisites of the deformability of the erythrocytes. The tanktread like rotation of the membrane around the cell content is merely a special form of continuous membrane deformation. Simultaneously, shear stresses are transmitted into the cell interior, a system of laminar shear is established here, and the whole cell participates in flow. In obviating DINTENFASS'ES [43] assumption of a "shear dependent liquefaction of membrane" for the explanation of the "low internal viscosity of the erythrocyte", these latter observations and measurements clearly establish that the red cell is able to participate in flow much like a fluid drop. The mechanism of continuous rotational deformation is not in conflict with the established biophysical characteristics of the cell content and the cell membrane. The viscosity of the cell content is likely to be similar to the exceptionally low viscosity of concentrated hemoglobin solutions (see p. 181). RAND and BURTON [128] as well as KATCHALSKY et al. [90] have shown that the membrane possesses a small flexion-rigidity and a high tensile rigidity and that the mechanical membrane properties are identical in all parts of the membrane [127, 128]. Whatever the ultimate details of the chemical, structural and micromechanical properties of the membrane will be, the flexibility is—at least in part—explained by the fact that the cell membrane is a thin shell [54].

The characteristic biconcave shape of the erythrocyte has given rise to numerous speculations relating to its functional significance in physiology [170] and hematology [117]. Most morphologists no longer assume that this shape is caused by a intracellular stroma [2]. FUNG has postulated that in the absence of pressure differentials, the membrane itself must be responsible for the natural shape of the erythrocytes: "the cell taking up the biconcave shape whenever it is completely relaxed for much the same reason as a rubber-glove has a natural shape" (FUNG).

Even though an uneven distribution of cholesterin on the different parts of the cell surface (MURPHY [117]) or the reversible formation of links between the two concavities of the erythrocyte at rest (Burton [22]) cannot be denied conclusively from these results, it is now generally accepted that the cell is always subjected to deformation when flowing. The multiple deformations of the erythrocytes during their passage of the capillaries of 5–7 μm diameter is a well documented phenomenon [16, 75, 154, 170, 180]. Many other kinds of deformations have been shown; Goldsmith [65], Monro [115] and Bloch [13] have presented evidence that the cells are also being continuously deformed into irregular polyhedra during flow in larger vessels. The shapes reported by these authors are very similar to those obtained in viscometric flow in the transparent cone plate viscometer (see p. 165, 182). Lastly, the regular ellipsoid deformations of the cells observed at either very high shear in plasma (see p. 165) or in viscous media are additional evidence suggesting that the factors and forces responsible for biconcavity are easily overcome by shearing forces in flow. This typical fluid drop like deformation and viscous adaptation supplies substantial experimental evidence supporting Fung's concepts. He regards the red cell as a membrane shell filled incompletely with a viscous incompressible fluid [54].

Whatever the ultimate structure of the membrane or consistency of the cell content in situ, the most important single factor facilitating deformation is the biconcavity. Due to the biconcave resting shape of the cell, the surface area to volume ratio is not minimized. As discussed in detail by Fung [54] such a body can be deformed into an infinite variety of shapes without changes in volume or surface area. For this, Fung used the term "isochoric applicable deformation". Provided that the content of the cell is liquid (in the sense that it cannot resist shear stresses without flowing) membrane rotation must be accompanying the observed ellipsoid transformation. The viscometric results obtained in red cell dextran dispersion can only be explained by assuming liquid cell interior participating in flow. By the same token, Goldsmith's [63] discovery that only fluid drops exhibit the strong axial migration so typical for red cells favors the concept of cell fluidity.

When a composite viscoelastic structure is subjected to external forces, for example the shear stress gradient by the surrounding fluid, the most rigid component is likely to resist the load [54]. In the case of either a solid interior or a non-bendable membrane, the stresses will cause the cell to rotate. Such rotation of non-deformable cells have indeed been observed in the present experiments and have been analyzed in detail by Goldsmith and Mason [63]. If, however, the bending rigidity is much smaller than the extensional rigidity, and the cell content is liquid, the tangential stresses imposed by the viscous-drag of the surrounding fluid will be taken up by the tensile element, resulting in deformation of the cell into an ellipsoid accompanied by continuous membrane rotation. The use of carbon tracers has supported this concept. Even though the limitations of light microscopy at these magnifications allow no unequivocal decision whether these carbon particles were actually attached to the membrane itself or only flowing in closely neighboring fluid layers, the existence of tracer rotations around a steadily oriented cell can only be explained by rotational motions of the membrane.

The shear induced membrane rotation is likely to transmit shear stresses into the interior of the cell, subjecting it to a system of laminar shear. Although this inner circulation has not yet been demonstrated, it must be deduced from the viscometric results obtained under experimental conditions identical to the microscopic observation of cell deformation. TAYLOR [160] and later RUM-SCHEIDT and MASON [136, 137] have demonstrated internal liquid circulation accompanying orientation and deformation. If one compares the viscosity of rigid cell and deformable cell suspensions, the effect of progressive recruitment of erythrocytes into fluidity becomes very evident.

Non-deformable red cells have an identical effect upon the suspension viscosity at high and low rates of shear. Deformable red cells even at the lowest shear rates have a smaller effect which diminishes with increasing shear rate. The flow curves of Fig. 26 demonstrate very clearly how anomalous viscosity of red cell dispersions is in fact the consequence of shear thinning caused by red cell fluidity. This is a very strong argument favoring the hypothesis that the difference observed between the viscosity of normal and rigid red cells in plasma (see p. 176) is also an indication of cell fluidity.

Following these results and those obtained with packed cells, a number of results seen in normal blood (see p. 164, 165, 170) can now be evaluated. It must be concluded, that the ellipsoid deformation of erythrocytes at 460 sec^{-1} (see p. 165), the irregular deformation of cells suspended in a ghost-plasma mixture and the progressive decrease in apparent blood viscosity after aggregate dispersion are evidence of a similar rotational deformation of red cells in normal blood at high rates of shear. The hypothesis is thus presented that in normal blood the fluid drop like adaptation of erythrocytes subjected to high shear is the common cause of cell deformation, shear thinning after aggregate dispersion, and the low apparent blood viscosity at rapid flow. Parallel to the fluid drop like transition, the blood adopts the rheological properties of an emulsion, i.e. a dispersion of fluid drops in another fluid.

VIII. Variations in Red Cell Rheology and their Influence on Blood Viscosity

The experiments reported thus far can best be summarized by the hypothesis that the erythrocyte suspended in plasma possesses bipotential flow properties. Subjected to low shear stresses, it is part of a three dimensional cell continuum (rouleaux network) which is able to resist these forces. Merely in response to high shear stresses, the cells become dispersed and acquire fluidity, thereby reducing the resistance to flow to a minimum unequalled in non-biological suspensions. The well-known shear rate dependence of human blood viscosity reflects these bipotential flow properties, apparent viscosity falling from infinity at stasis to a minimum at rapid flow. Before discussing the possible hemodynamic significance of this variability of viscosity (see p. 206) it must be stressed that the analysis of the shear-rate/shear stress data or pressure flow data from viscometers are not disclosing more than the bulk flow behavior of the blood under the shearing conditions prevailing in the particular experiment. Quite significant deviations in the flow behavior of the red cells and/or the plasma may be masked by apparently normal or only insignificantly altered bulk viscosity of blood. According to the present state of knowledge, a more detailed analysis is therefore necessary including the quantification of hematocrit, plasma viscosity, apparent viscosity of bulk blood at a wide range of shear rates, the aggregating potentials of the plasma, red cell aggregation and red cell deformability (fluidity) and the transition point (maximum shear rate of aggregation) between the flow of cell aggregates dispersion of aggregates and deformation of the individual erythrocytes.

The physiological fluidity of the erythrocyte, its most important flow facilitating property, may be reduced or abolished by a number of different factors: by changes in 1. the cell content and 2. the cell membrane and 3. by abnormal interaction between content and membrane or by immobilization of the erythrocyte in 4. an aggregate or 5. an agglutinate.

Rheology of Blood in Sickle Cell Disease

Collection of a rheological data under various flow conditions yields a much clearer picture about red cell rheology and its variation in disease than viscometry alone. This argument can be illustrated by studies executed with blood from patients suffering from homozyous sickle cell anemia, a disease in which rheological abnormalition have long been suspected (e.g. 42, 70, 121). The pO_2 dependence of the hemoglobin viscosity [121] offers a convenient tool to test its influence upon the flow behavior of the cell suspensions.

The hematocrit, the whole blood viscosity, the plasma viscosity and the filtrability of the venous blood of sickle cell patients showed greater variability than normal controls. Generally, the filtrability was less than in normal controls, and the apparent viscosity was abnormally high for the given hematocrit. As shown in Fig. 27, the viscosity of a 25 % sample at 230 sec^{-1} was 4 cP, or as high as a normal control with 40 % hematocrit (see p. 166). After correction for a hematocrit of 40 %, this increase in viscosity was even more pronounced. Packed cell viscosity at 92 % hematocrit was also considerably higher indicating reduced cell deformability. Upon lowering of the p_{O_2} to 51 mm Hg, the cell filtrability was completely abolished; the packed cells were so viscous that measurement became impossible, and the viscosity of the

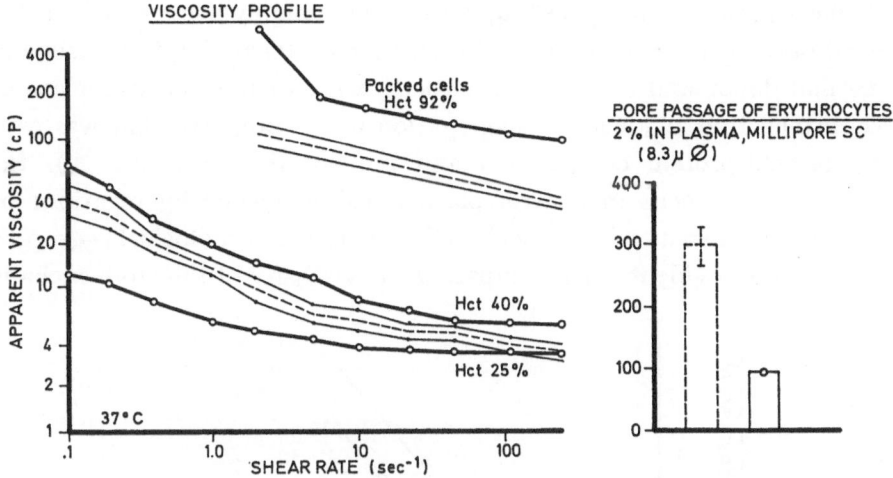

Fig. 27. Blood rheology in sickle cell anemia. Viscosity profiles (25%, 40% and 92% HCT) and red cell filterability. Normal controls are shown for comparison, details see text

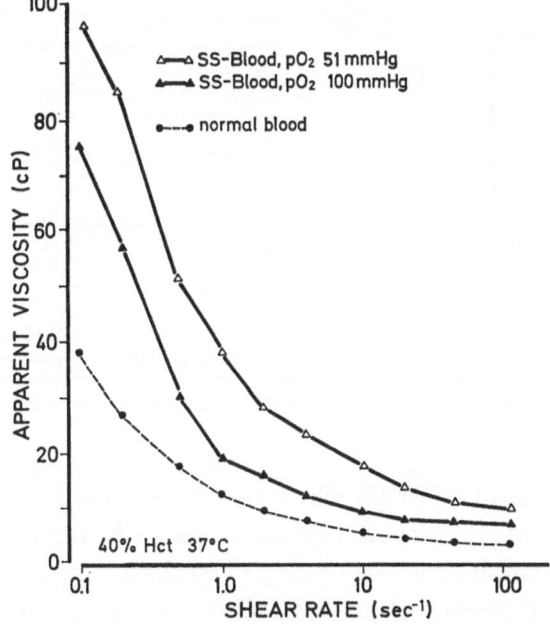

Fig. 28. Influence of pO_2 upon apparent viscosity of sickle cell blood. Note strong increase in high shear viscosity after moderate reduction of pO_2. Normal blood (unaffected by changes in pO_2) is shown for comparison

40% blood sample rose 11.2 cP at 230 sec⁻¹. This increase in viscosity at inter-mediate p_{O_2} was relatively less pronounced at low shear rates (Fig. 28). An-other abnormality found in this disease was a very pronounced tendency to aggregation at normal p_{O_2}; the cells formed aggregates which were similar to the ones described of the normal cells on p. 160. It was a conspicuous finding that in most patients the red cell aggregates were highly shear resistant, being dispersed only at a shear rate over 460 sec⁻¹. By exchanging the washed cells of a normal donor and a sickle cell patient, with their respective plasmas it was confirmed that tendency to aggregation was a property that was caused by the plasma-protein composition and not by the cells (Fig. 29). Both, normal and sickle cells in control plasma had a normal aggregation; both types of cells in plasma from the sickle cell patient had pathological aggregation. This became also evident when comparing the viscosity of the cross-exchanged

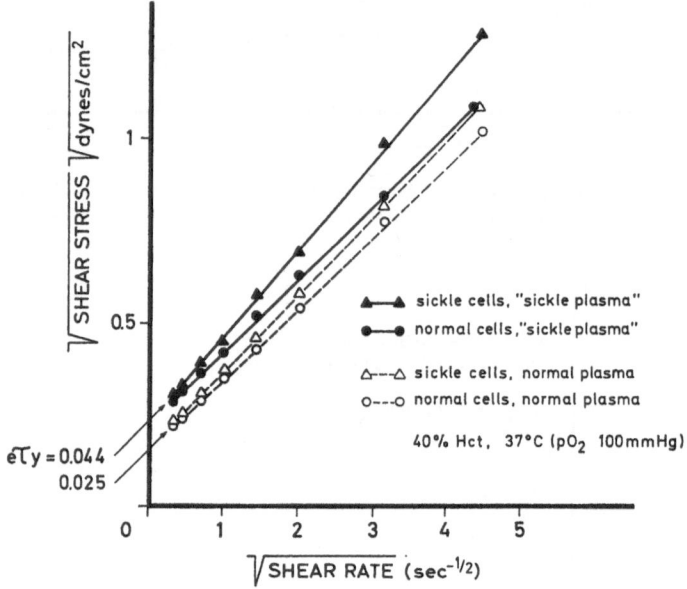

Fig. 29. Influence of plasmatic and cellular factors on rheology of sickle cell blood. Casson flow curves after cross exchanging of sickle cells, normal cells, plasma from sickle cell patients and normal plasma. Note the strong influence of sickle cell plasma on low shear viscosity (= shear stress for any given shear rate) and the strong influence of cellular factors on high shear viscosity

samples. At low rates of shear (in which flow is governed by aggregation), the samples containing sickle plasma had an abnormally high viscosity. At high rates of shear the samples containing sickle cells had a high viscosity. By using a Casson plot the influence of plasma factors and cell factors can be studied even better (Fig. 29). The samples containing sickle cells showed increased shear stress values at high rates of shear, the samples suspended in patients blood plasma had an increase in extrapolated yield shear stress. This finding was an experimental verification of the concept of bipotential flow properties of erythrocytes and changes thereof.

A third variable critical for abnormal flow behavior in sickle cell blood is the percentage of so-called irreversibly sickled cells (ISC), i.e. cells which do not retain their biconcave shape even at high p_{O_2}. By centrifuging a sample of sickle cell blood at $40.000 \times g$ for one hour, it became possible to obtain a sample of sickle cells practically free of (2.4%) ISC by taking the top layer of the cell column and resuspending it in plasma. Likewise, a sample rich in ISC was obtained by taking the bottom layer of cells (64% ISC). Fig. 30 shows the comparison of the two samples from the same patient: At high p_{O_2}, the sample poor in ISC shows practically normal viscosity and filterability, while the sample rich in ISC again demonstrates high viscosity indicating severed cell deformability. Lowering the p_{O_2} to 8 mm Hg again leads to marked in-

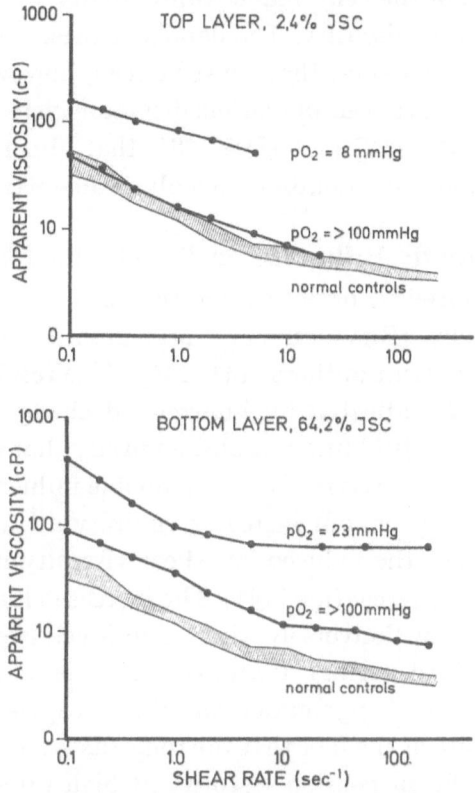

Fig. 30. Influence of the percentage of irreversibly sickled cells (ISC) and the pO_2 upon apparent viscosity of sickle cell suspensions. Note the normal rheology of ISC-poor sample at high pO_2 and the loss of shear thinning in the ISC-rich sample at low pO_2

creases in viscosity at high rates of shear. The sample rich in ISC showed almost Newtonian viscosity at low p_{O_2}: it behaved similar to the sample of the suspensions of aldehyde fixed cells in plasma. In both cases, the absence of cell fluidity did lead to a loss of the physiological shear thinning. This is an impressive evidence that the pronounced non-Newtonian viscosity of normal blood is caused by shear thinning due to transition of the cells into a fluid state. Shear thinning in the sample poor in ISC was restored after elevation of the p_{O_2}. These studies show how under physiological conditions loss of cell fluidity increases viscous resistance, especially in narrow tubes and in the living capillaries [70]. The possible role of these variables in the pathogenesis of acute crises and chronic degenerative alterations in sickle cell anemia merit further investigation. Similar alterations of red cells rheology have been reported by various authors [77, 78, 86, 87, 106, 118, 119, 162, 163, 169] in a great number of hemolytic anemias. In cases of acanthocytosis (a hemolytic anemia associated with a beta-lipoproteinemia; i.e. a defect in the lipid structure of the cell membrane) as well as in cases of pyruvate-kinase-deficiency (i.e. a defect in the enzyme system [106] of the cell), reduced filterability was seen combined with increases and high shear viscosity. The details of these studies are beyond the scope of the present discussion; they must be seen, however, in the context of the recent hypothesis advanced by various hematologists (e.g. Jandl [86, 87], Teitel [162], Murphy [118] and Ham [78]) that abnormal rheological properties are the common denominator of hemolytic anemias of different etiology.

Osmotic Influence on Blood Rheology

The changes produced by osmotic alteration of the red blood cells had been investigated repeatedly (Braasch [15], Meiselman [105], Murphy [118], Rand [109] and the present authors [143, 145]). The results obtained are very complex since they are affected by interrelated changes in hematocrit and osmolarity. Meiselman [105] first established firmly that for both swelling and crenation of the cells (hematocrit of 40%), a double influence is seen: crenation reduces viscosity near stasis, but increases viscosity at rapid flow. This result can now be understood: the reduced low shear viscosity is the consequence of disturbed or abolished aggregation [143]. The increased high shear viscosity reflects a loss in cell fluidity. Conversely, after osmotic cell swelling the aggregation behavior was maintained normal down to osmolarities of about 200 m Osmol: the maximum shear rate of aggregation and the extrapolated yield shear stress were unaffected in spite of the fact that the rouleaux were less flexible (Fig. 31). On the other hand, the suspension viscosity at high rates of shear was lower than that of normal controls at identical hematocrit as described by Rand [129] and Meiselman et al. [105]. A typical example is shown in Fig. 31. This decrease in high shear viscosity is likely to be caused by a reduction of the viscosity of the cell content. Under conditions in which cell deformation requires extensive membrane stresses (pore flow, see p. 177, flow of packed cell suspensions, see p. 179), the deformability of the erythrocyte is reduced, however. These studies therefore show that the viscosity of the intracellular hemoglobin solution does affect the viscosity of the blood in a fashion predicted by Taylor's equation. This can be demonstrated by measuring the viscosity of red cell dispersions under closely defined conditions of hematocrit (45%), plasma viscosity (1.2 cP), shear rate (230 sec^{-1}), and temperature (37° C). By plotting the relative viscosity (η_{susp}/η_0, Fig. 32) as a function of mean cellular hemoglobin concentration (MCHC), the general predictions of Taylor's

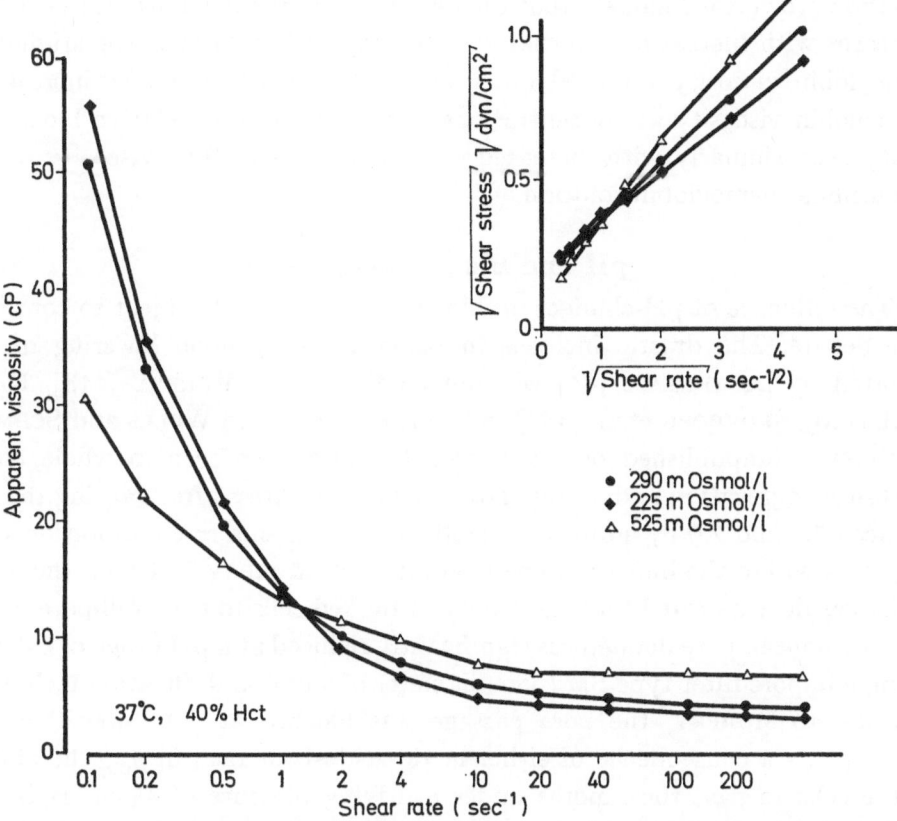

Fig. 31. Influence of osmotic swelling (225 mOsmol/litre) and osmotic crenation (522 mOsmol/litre) upon blood viscosity. Note the differing influence of either change upon high and low shear viscosity; for details see text

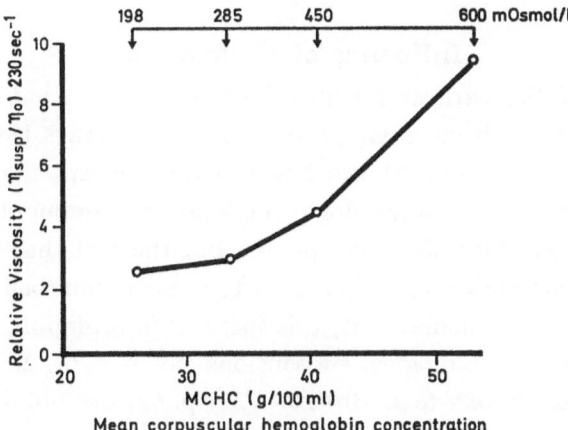

Fig. 32. Influence of osmotic swelling and osmotic crenation of red cells on viscosity of blood. Relative viscosity (η_{susp}/η_0) at 230 sec^{-1} rises as a function of mean cellular hemoglobin concentration. Hct = 45%. η_0 = 1.43 cP 37° C

equation are verified namely that the relative viscosity of emulsions (see p. 184) increases with increasing internal viscosity (η_1) of fluid drops. The studies on hemoglobin viscosity reported on p. 181 had established a steep increase in hemoglobin viscosity at concentrations above 32 g/100 ml, relative blood viscosity rises similarly with increased concentration and thus viscosity of the intracellular hemoglobin solution.

pH and Blood Rheology

The influence of pH-changes on blood viscosity is still subject to considerable debate. The drastic increase in blood viscosity upon lowering of pH reported by Dintenfass [44] was not confirmed by Murphy [118], Rand et al. [130], Skovborg et al. [155], Solvsteen [156], and Wells and Schmid-Schönbein (unpublished observations). The alteration seen in whole blood viscosity, aggregation and extrapolated yield shear stress after varying the pH between 7.8 and 7.0 by addition of lactic acid to plasma or equilibration with CO_2, was within the limits of experimental error. Murphy [118], on the other hand, has demonstrated that the ability of the red cells to pass Millipore filters of 8–9 μm mean pore diameter is significantly reduced at a pH range of 6.8–6.4. Using Millipore filter type SM (pore diameter of 5 μm) and filtration technique identical to Murphy, the pore passage was blocked at pH lower than 6.4 (Fig. 33). As a consequence of either lactic acidosis or rises in p_{CO_2} the ability of the cells to pass these pores under a driving pressure of 35 cm H_2O was abolished. It must be concluded therefore that bulk viscosity of while blood at pH levels compatible with life is unaffected by changes in pH. This notwithstanding, the flow behavior of cells subjected to the more severe changes in tissue pH found in shock, inflammation and burn may be significantly altered in their viscous behavior in the microcirculation, a phenomenon that merits further elucidation.

Influence of Hematocrit

It was one of the earliest results of hemorheology that blood viscosity increased markedly with increasing cell number (e.g. Hess [81], duPré et al. [124]). Although this is one of the few generally accepted rheological facts listed in most text books of physiology, the following comments are necessary: 1. not the increase in viscosity is surprising but the fact that blood is flowing at all at hematocrits above 60 %. Contrary to the common belief that viscosity rises steeply with rising hematocrit, this increase is negligible when compared to that of all other non-biological suspensions (see p. 176). 2. Viscosity values listed in most text books (e.g. Burton [22] p. 52) are obtained in capillary viscometers, i.e. under high shear rates. When measuring at lower rates of shear, rising hematocrit values lead to much more pronounced increases in apparent viscosity; 3. it is uncertain whether or not increases in hematocrit

Table 7. *Influence of changes in hematocrit on apparent blood viscosity (at various shear rates), on maximum shear rate of aggregation and on aggregate diameter (as measured in the rheoscope) 2 typical examples are shown*

	Hct	Shear rate 5.8 sec^{-1}		Shear rate 23 sec^{-1}		Shear rate 230 sec^{-1}		Maximum shear rate of aggregation
	%	η (cP)[a]	$\varnothing$ (μm)[b]	η (cP)[a]	$\varnothing$ (μm)[b]	η (cP)[a]	$\varnothing$ (μm)[b]	$\dot{\gamma}_{max}$ sec^{-1}
Normal	20	—	60	3.5	30	2.76	—	46
aggregation	43	9.8	90	7.8	30	4.41	—	46
	49	11.6	180	8.7	45	4.8	—	46
Intensified	10	—	90	2.6	60	1.95	15	230
aggregation	19	—	120	3.8	60	2.95	15	230
	40	12.0	150	8.9	90	5.20	15	230
	60	15.4	240	11.1	60	6.25	15	230

[a] apparent viscosity.
[b] aggregate diameter.

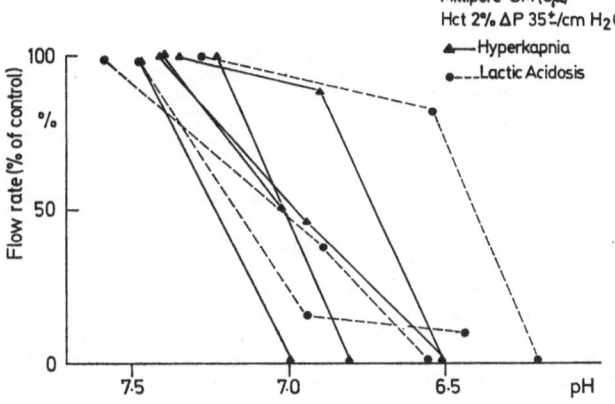

Fig. 33. Influence of lactic acidosis and hypercapnia on red cell filterability (flow rate in percent of control), Millipore type SM ΔP 35 cm H$_2$O. Note the pronounced decrease of red cell pore passage upon decrease of pH or increase of pCO$_2$

have the same influence on blood flow in vivo as they have in vitro (see p. 206). Blood viscosity near stasis is very strongly influenced by rises in hematocrit. In their initial study, MERRILL and co-workers [109] showed that the extrapolated yield shear stress rose as a function of the third power of the hematocrit. This led to the suspicion, that the process of aggregation per se was influenced. However, studies with the rheoscope showed that the maximum shear rate of aggregation ($\dot{\gamma}_{max}$, see p. 163) was unaffected by artificial changes in hematocrit between 10–60 %, the highest hematocrit in which plasma gaps and thus single aggregates could be discriminated. Table 7 shows two samples of normal blood and blood with pathological aggregation. The most pronounced change was that of aggregate size. This observation explained the apparent discrepancy between constant $\dot{\gamma}_{max}$ and variable yield shear stress. Not the process of rouleaux formation, but the consequences of the aggregation of rouleaux are influenced by hematocrit. The mechanical properties of the resulting three dimensional cell network are not only influenced by the adhesive links between cells, but also by the number of these links, the latter increases with increasing hematocrit. Therefore low shear viscosity and yield shear stress are markedly higher. This is illustrated by a study in which blood samples of 25, 50 and 75 % hematocrit from the same subject were compared. Deviating from the usual way of plotting, apparent viscosity is shown as a function of shear stress. The extrapolated yield shear stresses are included, as obtained by extrapolation

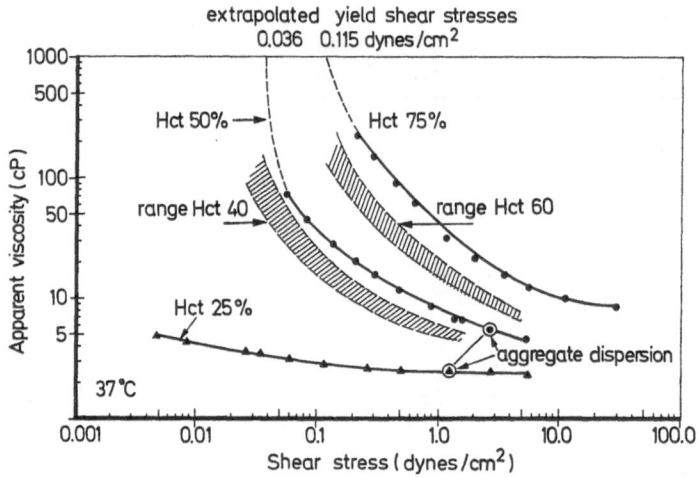

Fig. 34. Apparent viscosity of blood as a function of shear *stress* as influenced by changes in hematocrit. Data obtained by adjusting 25, 50 and 75 % HCT from the blood of a single donor. The hatched area shows the range ($x \pm$ S.D.) of viscosity data obtaned from the group of patients listed in Table 8. The γmax of the sample was 46 sec^{-1}, the shear stress values at this shear rate are encircled: the shear stress necessary to disperse aggregates in the 50 % sample is more than twice as high as in the 25 % sample

from a Casson plot of the same data (Fig. 34). The extreme shear rate dependence of high hematocrit blood is again seen. In comparing the 25 % and the 50 % sample at 0.02 dyne/cm² and at 2.0 dyne/cm², it becomes evident that under the conditions of viscometric flow at 0.02 dyne/cm², the 25 % sample is perfectly fluid (3.8 cP) while the 50 % sample behaves functionally like a solid—because this shear stress is lower than the extrapolated shear stress of the sample. At 2.0 dyne/cm², the viscosity of the 50 % sample (6.0 cP) is only about twice as high as the 25 % sample (2.45 cP), the difference decreasing with further increase in shear stress. The same point is illustrated in Table 8. Comparing the blood viscosities of 10 different blood donors at hematocrits of 40±1 % and 60±1 %, viscosity at 0.1–1.0 sec^{-1} is increased on average by 200 %, while at 20 sec^{-1} viscosity is increased by 100 %. From these studies it must be deduced that the combination of high hematocrit and low pressure gradients is likely to have a very pronounced hemodynamic effect.

Table 8. *Apparent viscosity and extrapolated yield shear stress of blood samples with a fixed hematocrit of 40 ± 1 and 60 ± 1 % ($\bar{x} + SD$). ($n = 10$)*

Shear rate	Hct 40% apparent viscosity ($\bar{x} \pm$ SD)	Hct 60% apparent viscosity ($\bar{x} \pm$ SD)
0.1 sec⁻¹	47.7 ± 15.2 cP	144.8 ± 37.2 cP
1.0 sec⁻¹	14.3 ± 4.66 cP	38.7 ± 11.2 cP
20.0 sec⁻¹	5.6 ± 0.25 cP	10.6 ± 1.7 cP
Extrapolated yield shear stress	0.027 ± 0.012 dyne/cm²	0.09 ± 0.032 dyne/cm²

Another point is illustrated in Fig. 34: Although the maximum *shear rate* of aggregation was identical in the 25 % and 50 % blood sample, the *shear stress* necessary to disperse the aggregates in the 50 % sample is markedly higher (2.7 vs·1.25 dynes/cm²) due to the overall higher viscosity level. This illustrates that the combination of high hematocrit and low shear stresses might lead to a viscious circle. Insufficient shear stresses permit the presence of aggregates, the aggregates increase viscosity and therefore lead to a slowing of the flow for the same shear stress, causing further aggregation and increased viscosity. Such aggregate-induced flow retardation may eventually lead to total stasis at which point blood behaves like a solid in the absence of coagulation.

Plasmatic Influences

A number of recent studies about rheological alterations in disease (e.g. [94, 157]) have established that significant changes occur in plasma viscosity (as high as 50 % above normal controls, and even higher in paraproteinemias). This is significant in itself by increasing hydraulic resistance. Besides, the changes in plasma composition underlying this increase in plasma viscosity (higher concentration of high molecular weight proteins) also affect red cell aggregation. The quantitative studies by MERRILL [108], WELLS [172], CHIEN [29] and SCHMID-SCHÖNBEIN [142] have shown that the physiological red cell aggregation is a consequence of an interaction of red cells and fibrinogen, leading to rouleaux formation and all its consequences. Later studies [175] in cases of pathologically intensified aggregation have shown that non-clottable serum proteins can cause red cell aggregation with all its consequences not merely in the classical cases of macroglobulinemia, but in diseases outside the field of hematology: for example in myocardial infarction and during recovery from surgical trauma. The details of pathological aggregation lead beyond the scope of the present study.

Influence of Plasma Ultrafiltration

The influence of plasmatic factors on the rheology of blood was also demonstrated in experiments in which the effect of capillary ultrafiltration was mimicked by concentrating the plasma proteins. Some rheological consequences of a 20% removal of plasma water (by dialyzing against vacuum) are illustrated in Fig. 35. The hematocrit is increased by 10%, the plasma protein concentration by 25%, the plasma viscosity (not shown) by 30%. All these changes lead to marked increases in blood viscosity, especially at low rates of shear. This is caused by an intensified red cell aggregation, the $\dot{\gamma}_{max}$ rising from 46 to 230 sec^{-1}, the extrapolated yield shear stress rising from 0.02–0.04 dyne/cm^2.

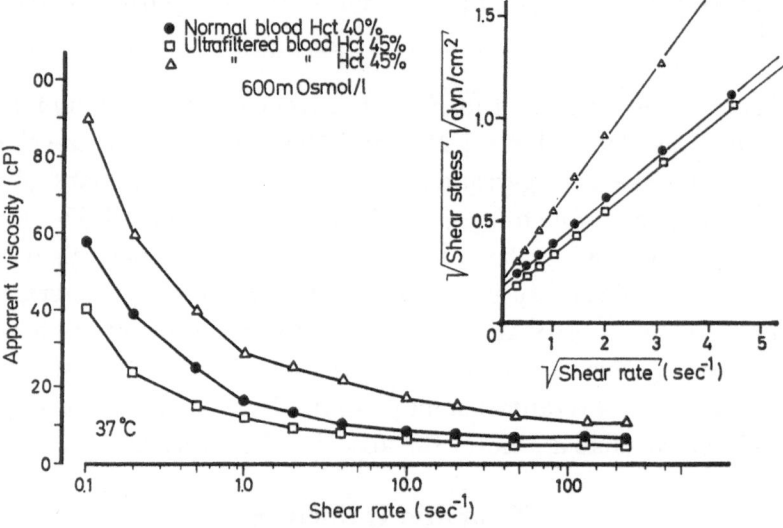

Fig. 35. Influence of ultrafiltration and hyperosmolarity on apparent viscosity of blood. A 20% filtration fraction leads to marked increases in apparent viscosity especially at low rates of shear. This is exaggerated after additional hyperosmolarity

These changes were even more pronounced if concentration of protein was combined with increases in osmolarity. In striking contrast to the behavior of crenated cells in normal plasma, where red cell aggregation is abolished or reduced (see p. 198), the crenated red cells in concentrated plasma formed very dense and rigid aggregates that were only dispersed at shear rates above 230 sec^{-1}. The reason for this behavior is obscure. At any rate the combination of cell rigidification (by osmotic dehydration) and red cell aggregation leads to pronounced increases in viscous resistance under all flow conditions examined. This viscous behavior might play a role in the hemodynamics of the vasa recta of the renal papilla, where high protein concentration after ultrafiltration and hyperosmolarity of the plasma are combined [146a].

Interaction of Plasma and Cells after Erythrocyte Dispersion

In light of red cell fluidity at high rates of shear, changes in relative viscosity (as defined on p. 152) are likely to be a function of the ratio of plasma viscosity and hemoglobin viscosity and/or overall cell deformability. The results of STONE et al. [158], who compared the viscous behavior of blood from various species, appear to confirm this hypothesis. It was found that those erythrocytes that were least deformable (as measured by packability under intermediate gravitational force) showed the highest suspension viscosity. At 50% hematocrit, goat cell suspensions were almost twice as viscous as human cell suspensions. ERSLEV and ATWATER [50] found that the viscosity as measured in an Ostwald-capillary viscometer rose as a function of mean cellular hemoglobin concentration. Their argument that this was caused by

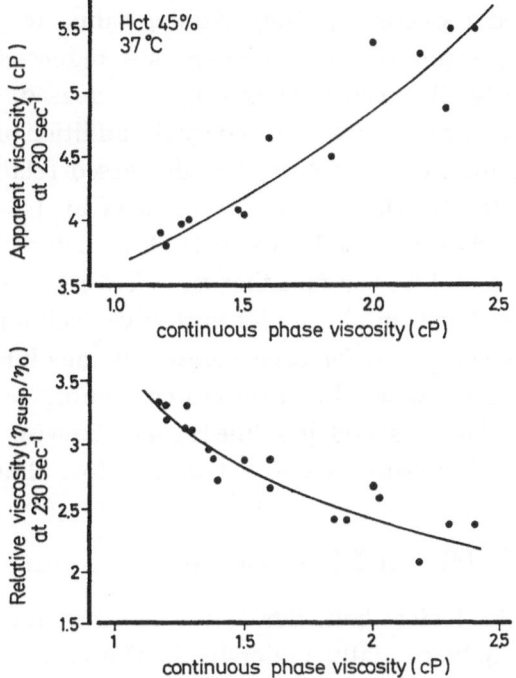

Fig. 36. Influence of changes in plasma viscosity [obtained by addition of saline and an isotonic solution of dextran 40000 (10%)] upon viscosity of red cell suspensions and relative viscosity (η_{susp}/η_o) (Hematocrit 45%, 37° C, 230 sec^{-1}). Note the decrease of relative viscosity with increasing plasma viscosity

increased "red cell viscosity" is convincing in light of the present studies. STRUMIA and PHILLIPS [159] have measured the influence of mean cellular volume on human blood viscosity, also using a capillary viscometer but expressing their data as relative viscosity (relative to H_2O). These authors have shown that blood viscosity is a function of the "relative volume of the red cell mass" (volume fraction) regardless of size and number of cells.

On the other hand, one can see in retrospect that after each increase in plasma viscosity, either by natural changes [48] or after admixture of Dextran [14, 18, 132, 179], the suspension viscosity was elevated less than the plasma viscosity, resulting in an actual reduction in relative viscosity (relative to plasma). This fall in relative viscosity has been attributed by Brooks [14] to changes in red cell charge. In repeating the experiments of Brooks et al. by suspending red cells in 10% dextran (M.W. 40 000), we observed typical fluid drop like deformation of the cells into ellipsoids. This makes it more likely that the reduction in relative viscosity that Brooks measured is a consequence of more effective cell fluidity by reducing the ratio of internal cell viscosity to plasma viscosity [150]. As predicted by Taylor's equation, each increase in plasma viscosity leads to a decrease in relative viscosity for the same hematocrit. Fig. 36 shows the relative viscosity of red cell suspensions under the same closely defined conditions described above (see p. 198). Adding saline to plasma reduces the viscosity of the suspension, but not as much as it reduces the viscosity of the plasma. Consequently, the relative viscosity is increased. On the other side, each increase in plasma viscosity (produced by the addition of dextran solutions) is accompanied by increased absolute, but decreased relative viscosity. This view is also supported by the data of Strumia et al. [159]. In their fig. 7, they plot "relative viscosity" (relative to H_2O) as a function of cell number and demonstrate the well known fact that the viscosity of red cells in saline is lower than that of red cells in plasma for each given cell number. If, however, relative viscosity is computed for each continuous medium used (see p. 152) an opposite result is obtained. For samples containing $4 \cdot 10^6$ RBC/mm^3, the relative viscosity in plasma is 2.08, in saline is 2.43. Likewise, at $6 \cdot 10^6$ RBC/mm^3 the relative viscosity in plasma is 3.36 versus 3.85 for saline.

IX. Blood Flow in the Vasculature

The details of rheological behavior of the blood in the living macro- and microcirculation is only partially understood. The classical work of Whittacker and Winton [178], confirmed by Levy and Share [98], has established that under normal perfusion conditions the effect of changes in hematocrit upon the viscous resistance in a living vascular tree is less than in costumary large bore viscometers. Even though a number of questions remain to be answered, e.g. kinetic energy correction (Burton [22], Benis [8]), the present knowledge suggests that the presence of red cells in plasma has a surprisingly small effect on viscous resistance in the living vasculature. Little can be added to Whittacker and Winton's comment to these phenomena after almost four decades: "The simplest explanations of the low values of apparent viscosity in the limb is based on the well known microscopic observation that corpuscles congregate near the axis of the tube, leaving a marginal zone of relatively clear plasma. Most of the relative movement of the concentric zones of liquid occur in this marginal zone, and hence the greater the proportion of the cross section of the tube occupied by the relatively free zone, the more closely the apparent viscosity of blood approaches that of plasma". He then reiterates Fahraeus

[53] conclusion that the "occupance of an almost solid rod of corpuscles passing down the axis of a tube at a higher velocity than that of the surrounding plasma" lowers the concentration of red cells actually present in a perfused vessel at any moment.

This reduction of the apparent hematocrit is based on the fact that the dynamic hematocrit is not the ratio of red cell volume to total volume (volume fraction), but the ratio of red cell flow rate to total flow rate (flow rate fraction; ZIERLER [182]). To ensure a certain constant volume fraction in the feed and recipient reservoirs, a much lesser dynamic hematocrit [99] in the connecting vessel is necessary provided the velocities of plasma and red cells differ considerably. Such velocity differences have actually been measured by THOMAS and ROWLANDS [165, 134] in entire organs.

While WHITTACKER's experiments were performed "without prejudice to the actual physical explanation of the variation in internal friction" [178], a number of recent studies have contributed towards its explanation. The deformation of the red cells in capillaries smaller than the resting diameter of the cells (GUEST [75], BRANEMARK [17]) as well as the dynamics of axial migration of red cells in flow (GOLDSMITH [63]) have been firmly established. It is generally accepted that both phenomena decrease the viscous resistance caused by blood flowing in small blood vessels. Both phenomena are the consequence of the responses of the deformable red cell to the forces acting upon it.

LIGHTHILL [101] has recently proposed the theory that the movement of blood cells through the very narrow capillaries has to be treated as a motion of solid pellets being forced along distensible tubes. The present knowledge about the physical characteristics of capillaries (FUNG) [54] and the rheological properties of the erythrocytes as described herein invalidate most of LIGHTHILL's basic assumptions about the physical properties of the components he analyzed.

Strictly speaking, the present hemorheological results are only pertinent to the specialized flow conditions of blood in the cone plate viscometer. However, their general significance is based on the fact that the factors found responsible for the anomalous viscosity of bulk blood are identical to those long observed in the microcirculation, namely red cell deformation and red cell aggregation. The most unique characteristic is red cell deformability (fluidity). Its consequence is the unusually low apparent viscosity of red cell suspensions, especially at high shear rates and hematocrits. The physical characteristics of the erythrocyte which allow fluid-drop-like transition of the cell are the very ones responsible for the great deformability in capillary flow (FUNG [54]) as well as for axial migration of the erythrocytes (GOLDSMITH [63]). The migration of the erythrocytes along shear stress gradients not only explains "plasma skimming" [96], but also redistribution of cells and plasma [12] and reduction of "apparent hematocrit" as established since FAHRAEUS [53]. It is obvious that this factor contributes to the decrease in viscosity with falling vessel radius.

This stands in contrast to the behavior of rigid particles that have been shown to plug flow in tapered vessels larger than their own diameters in concentrations over 20 % (Sacks and Tickner [138] [163 a]). The flow regime in arterioles and capillaries is characterized by high shear rates and high shear stresses under normal perfusion conditions. In spite of small actual velocities, the velocity gradients are high as a consequence of small vessel radius. No true measurements of shear stresses and shear rates have been reported in the literature for living tissues. Calculated values we obtained from applying an arbitrary flow rate to a vascular bed analysed by Wiedeman [176] (assuming continuity of perfusion) indicate maximum shear in arterioles and capillaries [146]. Sacks and Tickner [138] assumed a similar distribution. Goldsmith calculated optimum conditions for axial migration of red cells in vessels smaller than 500 μm [63].

The aggregation of red cells was observed to occur as a consequence of reduced shear rates in cone plate flow [141], capillary flow in vitro (Benis [6]), and in vivo (Schmid-Schönbein [146], Thuranskii [166] and Wells [173]). In large bore viscometers, aggregation is linked to increases in apparent viscosity near stasis. There are no data presently available to judge whether such an increase in viscosity also occurs in vitro in capillaries smaller than about 300 μm, although Benis's [4] data give a slight indication of this. The fall in apparent viscosity measured frequently (e.g. [3, 58]) in such small vessels (Fahraeus-Lindqvist effect [53]) only applies to the flow at high rates of shear. Intuitive reasoning suggests that at low shear rates the presence of rod-like particles should lead to even more pronounced hydraulic resistance in small vessels. It is also unknown whether or not such an increase in viscosity associated with low shear flow and red cell aggregation occurs in vivo [9]. It is very unlikely that this question can be answered adequately from the analysis of pressure-flow curves obtained in entire organs. The presently available data are equivocal [1, 39, 46]. The presence of a positive intercept in pressure-flow curves (zero flow at finite pressure) which is customarily related to critical closure of small blood vessels (Burton [21]) may at least in part be related to the rheological properties of red cell aggregates. Analysis of the flow dynamics in smaller compartments of the microcirculation or the sequential resistance along the vascular tree may yield results different from those obtained in entire organs. Data obtained by Baeckström [1] et al. indicate that the number of perfused vessels in hypotension is influenced by the rheological properties of the blood. The frequent observation of prolonged, but reversible stasis of aggregated blood in venular vessels in shock favors the assumption that the rheological properties of aggregated blood at low rates of shear may have an even more pronounced hemodynamic effect than can be deduced from measurements in vitro. The phenomenon of prolonged reversible stasis is likely to be due to factors of the blood itself. Although active vascular occlusion by vasomotor action may initiate this stagnation, it is unlikely to be sustained by active occlusion in the presence of the metabolic consequence of such stasis, namely anoxia, hypercapnia and acidosis causing vasodilatation.

The velocity profile of blood flowing in small blood vessels is also influenced by the increase of blood viscosity at low rates of shear (due to red cell aggregation). A blunting of the viscosity profile i.e. a deviation from the parabolic velocity distribution in small glass tubes perfused with blood have been reported by BUGLIARELLO [20], GAEHTGENS and WAYLAND [57], and GOLDSMITH [64]. BERMAN and FUHRO [10] reported blunting and the development of "plug flow" in the arterioles of the hamster cheek pouch. The flow of an unsheared core region takes place which is surrounded by a sleeve of plasma in which all shearing occurs. The transition from plug flow to sheared flow was shown to be a function of the shear rate at the wall and the vessel radius: the critical shear rate necessary to reduce sheared flow increased with decreasing vessel radius down to 10 μm. This measurement explains MONRO's [114] observation about cell aggregates in all arterioles of the rabbit ear chamber and renders debatable certain hypotheses about plugging of arterioles by red cell aggregates in cases of "blood sludging" (KNISELY [92], GELIN [57]).

All these results underline the concept (MÜLLER [116]) that the viscosity of blood cannot be treated as a physical constant in the customary sense. It must be cautioned against extrapolation from shear stress/shear rate and pressure/flow rate data to the hemodynamic significance of apparent blood viscosity under any other than the given flow conditions in viscometer. There is evidence suggesting that the viscosity of the blood flowing in the circulation can vary between that of plasma alone in a rapidly perfused capillary and infinity in any static vascular segment. PROTHERO and BURTON [126], GERBSTÄDT et al. [58], and BARRAS [3] have shown that the apparent viscosity of blood in very small capillaries may approach that of plasma. This phenomenon was predicted theoretically by WHITMORE [180], HOCHMUTH [83] and many others for a system in which the suspended particles diameter after deformation is smaller than the capillary diameter. In such systems, the frictional energy loss takes place in marginal layers. The presence of small, deformable or fluid particles in the axial portion of low shear has a small additional effect (LEW and FUNG [99], HOCHMUTH [83]).

This inability to define the value of the viscosity in the microvasculature notwithstanding, viscous resistance is the prime factor in the hemodynamic of the peripheral circulation. There have been attempts to separate conceptionally the "viscous resistance" caused by blood from the "vascular resistance". These attempts are misleading; the only reason for pressure drop across the resistance vessels is the fact that the blood is not an ideal fluid (in BERNOULLI's sense) but possesses some finite viscosity. However, from the data presently available about pressure drops in the circulation, we must assume that the viscous energy dissipation in aorta, arteries and arterioles down to about 40 μ is comparatively small (Burton [22], Zweifach [178]), probably due to the comparatively low flow velocities and consequently small shear rates. The ramification of the arterial tree into arterioles and capillaries leads to flow conditions in which the transport of a given blood volume requires much higher energy dissipation. Our present knowledge about blood flow in small vessels in vitro [3, 58, 126] and in vivo suggests, however, that during rapid flow through these vessels the apparent viscosity of the blood falls to a level close to that of plasma.

The pressure energy supplied to the blood by the heart is thus primarily dissipated during blood flow in the terminal branches of the arterial tree. Due to their abundance of mural smooth muscle and their small diameter, these small arterial vessels can also greatly change their diameter and thence regulate actively the rate of energy dissipation and consequently flow. The steep pressure drop across the "resistance vessels" is, of course, regulated by the vascular tone [183]. However, it would be much steeper if the viscous resistance offered by the blood would be equal to that of blood in large tubes. Besides, the steep pressure gradient across small arteries and arterioles results in high wall shear stresses and thereby facilitates axial migration (Goldsmith [63]). This is a field that offers formidable experimental and mathematical tasks for future research.

The hemodynamic significance of the great variability in apparent blood viscosity in the microcirculation is underlined by the fact that this part of the vasculature makes up the largest part of the volume capacity of the vascular bed. Knisely [93] has shown that in the rat more than 80% of the vascular volume is taken up by vessels smaller than 180 μm in diameter. Very little is known about the distribution of this vast capacity among the different vessels of the microcirculation. Wiedeman's [176] measurements in the living micro-vascular bed of the bat wing indicate that more than 80% of this vascular volume can be taken up by venules and veins smaller than 76 μm in diameter, a finding also reported by Litton and Berman [102] for the hamster cheek pouch. These findings vary considerably from those data extrapolated from measurements of Mall [103] on histological sections of dog intestine, quoted in most textbook of physiology.

In summary, the discovery of red cell fluidity allows for the first time a unifying interpretation of the data presently available about the flow behavior of blood in both micro- and macrocirculation. Reviewing the recent micro-rheological results obtained by various laboratories (Chien, Dintenfass, Goldsmith, Schmid-Schönbein), Chien [33]) has recently published a unifying explanation for the shear dependence of blood viscosity in viscometric flow. He postulated that "red blood cell aggregation and red blood cell deformation may exert their rheological effects by a common mechanism, namely shear dependent changes in the effective cell volume". According to this hypothesis, red cell aggregates immobilize plasma and therefore increase the effective red cell volume. Conversely, the alignment of deformed red cells reduces the volume of immobilized external fluid and decreases the disturbance of external stream lines. CHIEN argued in essence, that the different apparent viscosities at low and high rates of shear were a function of the disturbance of red cells to the external stream lines in the plasma. This lucid interpretation can be generalized to explain the hemodynamic significance of red cell fluidity in both the macro- and microcirculation. The rheological behavior of dispersions can be analyzed in terms the disturbance that the dispersed phase (be it solid or fluid) causes to the flow of the continuous phase in the perfused vessels. As the consequence of its fluidity, the disturbance that the red cells render to the flow of plasma through the vascular tree is reduced to a minimum. In large bore vessels, the viscous behavior of the blood as an emulsion causes much less resistance than that of suspensions would (see p. 176). In arterioles, the fluidity reduces viscosity, but in addition causes the radial migration of the red cells towards the vessels axis. This moves the red cells to a rapidly flowing axial plug of red cells in which no shear takes place and which is surrounded by a lubricating layer of plasma. Since the energy dissipation in the rapidly moving axial part of perfused vessel is comparatively small (even for pure plasma, due to the minute velocity gradient), the additional pressure drop caused by the presence of erythrocytes appears to be minimized. The same argument holds true for the perfusion of true capillaries; here, too, the deformed and fluid erythrocytes are moving in an axial train (WHITMORE [180]) surrounded by a lubricating plasma layer. This adaptation of the erythrocytes not only allows blood flow in vessels smaller than the resting diameter of the red cell, but allows blood flow at the lowest possible viscous resistance, namely that of plasma itself.

The unique fluidity of the erythrocytes can be impaired by a variety of factors, e.g. by red cell aggregation and by red cell agglutination, by membrane or by hemoglobin defects, or by a reduction of the shear stresses in the micro-circulation. All these conditions have in common that flow characteristics of blood change from those of an emulsion to those of a suspension. This change results in a concomitant increase in the disturbance of plasma flow caused by the erythrocytes.

The cause of impaired erythrocyte fluidity can be determined using the methods developped in the present study (red cell filtration, packed cell viscometry, rheoscopy). The consequences to the flow of blood under various shear stresses can be assessed in vitro by measuring the apparent viscosity, or better the apparent relative viscosity (s. p. 152) of blood. Immobilisation of erythrocytes exerts rheological effects primarily at low rates of shear, while loss of the individual cell's fluidity increases relative viscosity at high rates of shear.

References

1. Baeckström, P., Folkow, B., Löfving, B., Kovach, A. B. G., Öberg, B.: Evidence of plugging of the microcirculation following acute hemorrhage. Proc. VI. Conf. on Microcirculation (Aalborg 1970). Bibl. anat. (Basel) 11 (in press).
2. Baker, R. F.: Ultrastructure of the red blood cell. Fed. Proc. 26, 1785–1801 (1967).
3. Barras, J. P.: L'écoulement du sang dans les capillaires. Helv. med. Acta 34, 468–477 (1968).
4. Bayliss, L. E.: Rheology of blood and lymph. In: Deformation and flow in biological systems (A. Frey-Wissling, ed.), p. 354–418. Amsterdam: North Holland Publ. Co. 1952.
5. — The rheology of blood. In: Handbook of Physiology, Sect. 2, Circulat. (I. W. F. Hamilton and P. Dow, eds.), p. 137–150. Washington, D.C. 1962.
6. Benis, A. M.: The flow of blood through models of the microcirculation. Doctorate thesis. Massachusetts Institute of Technology (1964).
7. — Lacoste, J.: Study of erythrocyte aggregation by blood viscometry at low shear rates using a balance method. Circulat. Res. 22, 29–41 (1968).
8. — Usami, S., Chien, S.: Role of viscosity and inertial losses in pressure-flow relations studied on perfused canine hindpaw. Fed. Proc. 29, 259 Abs. (1970).
9. Bermann, H. J.: Rheological properties of the microvasculature. Bibl. anat. (Basel) 7, 29–34 (1965).
10. — Fuhro, R. L.: Effect of rate of shear on the velocity-profile and orientation of red cell in arterioles. Bibl. anat. (Basel) 10, 32–37 (1969).
11. Bingham, E. C., Roepke, R. R.: The rheology of blood. IV. The fluidity of whole blood at 37° C. J. gen. Physiol. 28, 131–149 (1944).
12. Bloch, E. H.: A quantitative study of the hemodynamics in the living vascular system. Amer. J. Anat. 110, 112–154 (1962).
13. — High speed cinephotography of the microvascular system. In: Hemorheology, Proc. 1st Intern. Conf. (A. L. Copley, ed.), p. 655–667. Oxford: Pergamon Press 1968.
14. Bollinger, A., Simon, H. J., Köhler, R., Lüthy, E.: Wirkung von niedermolekularem Dextran auf Blutviskosität und Extremitätendurchblutung. Z. Kreisl.-Forsch. 57, 456–465 (1968).
15. Braasch, D.: Verminderte Erythrozytenflexibilität (hervorgerufen durch Barbiturate, Verbrennung, Hypoxämien) und ihre Wirkung auf den Kapillarkreislauf. Pflügers Arch. ges. Physiol. 278, 130–140 (1963).
16. Branemark, P. I., Lindström, J.: Shape of circulating blood corpuscles. Biorheology 1, 139–142 (1963).
17. Brooks, D. E., Goodwin, J. W., Seaman, G. V. F.: Interactions among erythrocytes under shear. J. Appl. Physiol. 28, 172–177 (1970).
18. Brundage, J. T.: Blood and plasma viscosity determined by the method of concentric cylinders. Amer. J. Physiol. 110, 659–665 (1934).
19. Bugliarello, G., Hayden, G.: Detailed characteristics of the flow of blood in vitro. Trans. Soc. Rheol. 7, 209–230 (1963).
20. Burton, A. C.: The physical equilibrium of small blood vessels. Amer. J. Physiol. 149, 389–399 (1947).
21. — Physiology and biophysics of the circulation. Chicago: Year Book Medical Publ. Inc. 1965.
22. — In: Ciba Symposion on Mass Transport. London: Churchill 1968.
23. Canham, P. B., Burton, A. C.: Distribution in size and shape in populations of human red cells. Circulat. Res. 22, 405–417 (1968).
24. Casson, N.: A flow equation for pigment-oil suspensions of the printing ink type. In: Rheology of disperse systems (C. C. Mill, ed.). New York-London-Paris-Los Angeles: Pergamon Press 1959.
25. Charm, S. E., Kurland, G. S., Brown, S. L.: The flow characteristics of blood suspensions. In: Biomedical fluid mechanics Symposion, Denver 1966; p. 89–93. New York: Amer. Soc. Mech. Eng. 1966.

26. CHARM, S. E.: Static method for determining blood yield stress. Nature (Lond.) **216**, 1121–1123 (1967).

27. — Discrepancy in measuring blood in couette, cone plate and capillary tube viscometers. J. appl. Physiol. **25**, 786–789 (1968).

28. CHIEN, S., DELLENBACK, R. J., USAMI, S., GREGERSEN, M. I.: Plasma trapping in hematocrit determination. Difference among animal species. Proc. Soc. exp. Biol. (N.Y.) **119**, 1155–1161 (1965).

29. — USAMI, S., TAYLOR, H. M., LUNDBERG, J. L., GREGERSEN, M. I.: Effect of hematocrit and plasmaproteins on human blood rheology at low rates of shear. J. appl. Physiol. **21**, 81–87 (1966).

30. — — DELLENBACK, R. J., GREGERSEN, M. I.: Blood viscosity: Influence of erythrocyte deformation. Science **157**, 827–829 (1967).

31. — — — — NANNINGA, L. B., GUEST, M. M.: Blood viscosity: Influence of erythrocyte aggregation. Science **157**, 829–831 (1967).

32. — DELLENBACK, R. J., USAMI, S., SEAMAN, G. V. F., GREGERSEN, M. I.: Centrifugal packing of suspensions of erythrocytes hardened with acetaldehyde. Proc. Soc. exp. Biol. (N.Y.) **127**, 982–985 (1968).

33. — Shear dependence of effective cell volume as a determinant of blood viscosity. Science **168**, 977–979 (1970).

34. COLEMANS, B. D., MARKOVITZ, H., NOLL, H.: Viscometric flows of non-Newtonian fluids. Berlin-Heidelberg-New York: Springer 1966.

35. COKELET, G. R., MERRILL, E. W., GILLILAND, E. R., SHIN, H.: The rheology of human blood. Measurement near and at zero shear rate. Trans. Soc. Rheol. **7**, 303—317 (1963).

36. — The rheology of human blood. D. Sc. Thesis, Massachusetts Institute of Technology (1963).

37. — MEISELMAN, H. J.: Rheological comparison of hemoglobin solutions and erythrocyte suspensions. Science **162**, 275–277 (1968).

38. COUETTE, M. M.: Ann. Chim. Phys. **21**, 433 (1890). Cit. bei WAYLAND (Ref. 168).

39. DICKMANS, H. A., GAETHGENS, P. A. L., EISOLT, J., HIRSCH, H.: The effect of Dextran-induced changes in the flow properties of blood upon resistance to flow in the intestinal vascular bed. Proc. VI. Conf. on microcirculation (Aalborg 1970). Bibl. anat. (Basel) **11** (in press).

40. DINTENFASS, L.: Thixotropy of blood and proneness to thrombus formation. Circulat. Res. **11**, 233–239 (1962).

41. — Viscosity and clotting of blood in venous thrombosis and coronary occlusion. Circulat. Res. **14**, 1 (1964).

42. — Rheology of packed red cells containing hemoglobins AA, S-A and S-S. J. Lab. clin. Med. **64**, 594–600 (1964).

43. — Molecular and rheological considerations of the red cell membrane in view of the internal fluidity of the red cell. Acta haemat. (Basel) **32**, 299–313 (1964).

44. — BURNARD, E. D.: Effect of hydrogen ion concentration on the in-vitro viscosity of packed red cells and blood at high hematocrits. Med. J. Aust. **1**, 1072–1074 (1966).

45. — Internal viscosity of the red cell and a blood viscosity equation. Nature (Lond.) **219**, 956–958 (1968).

46. DJOJOSUGITO, A. M., FOLKOW, B., ÖBERG, B., WHITE, S.: A comparison of blood viscosity measured in vitro and in a vascular bed. Acta physiol. scand. **78**, 70–84 (1970).

47. DRABKIN, D. L., AUSTIN, J. H.: Biol. Chem. **112**, 51 (1935).

48. EHRLY, A. M.: Hämorheologische Probleme bei Venenerkrankungen. Zbl. Phlebologie **6**, 338–343 (1967).

49. EINSTEIN, A.: Eine neue Bestimmung der Moleküldimensionen. Ann. Physik **19**, 289–306 (1906).

50. Erslev, A. J., Atwater, J.: Effect of mean corpuscular hemoglobin concentration on viscosity. J. Lab. clin. Med. **62**, 401–406 (1963).
51. Fahraeus, R.: Suspension-stability of the blood. Acta med. scand. **55**, 1–228 (1921).
52. — The suspension stability of blood. Physiol. Rev. **9**, 241–274 (1929).
53. Lindqvist, T.: The viscosity of blood in narrow capillary tubes. Amer. J. Physiol. **96**, 562 (1931).
54. Fung, Y. C.: Theoretical considerations of the elasticity of red cells and small blood vessels. Fed. Proc. **25**, 1761–1772 (1966).
55. — Blood flow in the capillary bed. Biomechanics **2**, 353–372 (1969).
56. Gaehtgens, P., Meiselman, H. J., Wayland, H.: Velocity profiles of human blood at normal and reduced hematocrit in glass tubes up to 130 μ m diameter. Microvascular Res. **2**, 13–23 (1970).
57. Gelin, L. E., Zederfeldt, B.: Experimental evidence of the significance of disturbances in the flow properties of blood. Acta chir. scand. **122**, 336–342 (1961).
58. Gerbstädt, H., Vogtmann, C. H., Rüth, P., Schöntube, E.: Die Scheinviskosität von Blut in Glaskapillaren kleinster Durchmesser. Naturwissenschaften **53**, 526 (1966).
59. Gillison, P. J., Dauwalter, C. R., Merrill, E. W.: A rotational viscometer using A.C. torque to balance loop and air-bearing. Trans. Soc. Rheol. **7**, 319–331 (1963).
60. Goldsmith, H. L., Mason, S. G.: Some model experiments in haemodynamics. Bibl. anat. (Basel) **4**, 462–478 (1964).
61. — — Some model experiments in haemodynamics II. Bibl. anat. (Basel) **7**, 353–362 (1965).
62. — Microscopic flow properties of red cells. Fed. Proc. **26**, 1813–1820 (1967).
63. — Mason, S. G.: The microrheology of dispersions. In: Rheology, theory and applications, vol. IV (F. R. Eirich, ed.), p. 86–249. New York: Academic Press 1967.
64. — The microrheology of red blood cell suspensions. J. gen. Physiol. **52**, 5–27 (1968).
65. — Mason, S. G.: Model particles and red cells in flowing concentrated suspensions. Bibl. anat. (Basel) **10**, 1–8 (1969).
66. — Flow and deformation of red blood cells in concentrated suspensions. Fed. Proc. **28**, 423 (1969).
67. — Beitel, L.: Axial migration of red cells in tube flow. Fed. Proc. **29**, 319 (1970).
68. Goldstone, J., Hutchins, P. M., Wells, R. E.: In vivo flow behavior of non-deformable erythrocytes. Microvascular Res. **2**, 1 (1970).
69. — Schmid-Schönbein, H., Wells, R. E.: The rheology of red blood cell aggregates. Microvascular Res. **2**, 273–786 (1970).
70. — Hutchins, P. M., Mentzer, W. C., Nathan, D., Wells, R. E.: Erythrocyte deformation in sickle cell disease. Proc. XIII. Internat. Congr. Hematology, Munich 1970.
71. Goll, K. H.: Vergleich der Methoden der Erythrozytensedimentationsgeschwindigkeit. Acta biol. med. germ. **2**, 590–598 (1959).
72. Gregersen, M. I., Chien, S., Peric, B., Taylor, H.: Investigations of blood viscosity at low rates of shear: effects of variations in the concentration and character of the red cells and in the composition of the suspending medium. Bibl. anat. (Basel) **7**, 383–384 (1965).
73. — Bryant, C. A., Hammerle, W. E., Usami, S., Chien, S.: Flow characteristics of human erythrocytes through polycarbonate sieves. Science **157**, 825–827 (1967).
74. — — Evaluation of deformability of red cells by sieving tests. Hemorheology, Proc. 1st Intern. Conf. Reykjavik 1966 (A. L. Copley, ed.), p. 539–549. Oxford: Pergamon Press 1968.
75. Guest, M. M., Bond, T., Cooper, R. G., Derrick, J. R.: Red blood cells: change in shape in capillaries. Science **142**, 1319–1320 (1963).
76. Hagen, G.: Über die Bewegung des Wassers in engen zylindrischen Röhren. Poggendorf'sche Ann. Phys. Chem. **46**, 423–442 (1839).

77. HAM, T. H., DUNN, R. F., SAYRE, R. W., MURPHY, J. R.: Physical properties of red cells as related to the effects in vivo. I. Increased rigidity of erythrocytes as measured by viscosity of cells altered by chemical fixation, sickling and hypertonicity. Blood **32**, 847–861 (1968).

78. HAM, T. H., SAYRE, R. W., DUNN, R. F., MURPHY, J. R.: Physical properties of red cells as related to effects in vivo. Effect of thermal treatment on rigidity of red cells, stroma and sickle cells. Blood **32**, 862–871 (1968).

79. HAYNES, R. H., BURTON, A. C.: Role of the non-Newtonian behavior of blood in hemodynamics. Amer. J. Physiol. **197**, 943–950 (1959).

80. — Physical basis of the dependence of blood viscosity on tube radius. Amer. J. Physiol. **198**, 1193–1200 (1960).

81. HESS, R. W.: Blutviskosität und Blutkörperchen. Pflügers Arch. ges. Physiol. **140**, 354–362 (1911).

82. — Gehorcht das Blut dem allgemeinen Strömungsgesetz der Flüssigkeiten? Pflügers Arch. ges. Physiol. **162**, 187–244 (1915).

83. HOCHMUTH, R. M., SUTERA, S. P.: Large scale model studies of apparent viscosity and erythrocyte velocity in capillaries. Bibl. anat. (Basel) **10**, 113–123 (1969).

84. JACOBS, H. R.: The "viscosity" of red cell packs. Biorheology **1**, 129–138 (1963).

85. — The deformability of red cell packs. Biorheology **1**, 233–238 (1963).

86. JANDL, J. H., SIMMONS, R. L., CASTLE, W. B.: Red cell filtration and the pathogenesis of certain hemolytic anemias. Blood **18**, 133–148 (1961).

87. — ASTER, R. H.: Increased splenic pooling and the pathogenesis of hypersplenism. Amer. J. med. Sci. **27**, 383–397 (1967).

88. JEFFERY, G. B.: The motion of ellipsoidal particles immersed in a viscous fluid. Proc. roy. Soc. A **102**, 162–179 (1922).

89. JOBLING, A., ROBERTS, J. E.: Some observations on dilatancy and thixotropy. In: Rheology of dispersed systems (C. C. MILL, ed.), p. 127–138. New York: Pergamon Press 1959.

90. KATCHALSKY, A., KEDEM, O., KLIBANSKY, C., DE VRIES, A.: Rheological considerations of the hemolysing red blood cell. In: Flow properties of blood and other biological systems (A. L. COPLEY and G. STAINSBY, eds.), p. 155–171. New York: Pergamon Press 1960.

91. KNISELY, M. H., BLOCH, E. H., ELIOT, T. S., WARNER, L.: Sludged blood. Science **106**, 431–433 (1947).

92. — Intravascular erythrocyte aggregation (blood sludge). In: Handbook of physiology, sect. 2, vol. III (W. F. HAMILTON and P. Dow, eds.), p. 2249–2292. Washington D.C. 1965.

93. KNISELY, W. H., MAHALEY, M. S., JETT, H. H.: Approximation of total vascular space and its distribution in three sizes of blood vessels by plastic casts. Circulat. Res. **6**, 20–25 (1958).

94. KOK, D. A.: Studies of the viscosity of serum in disease states, using a capillary viscosimeter. Biorheology **3**, 216–217 (1966).

95. KREUZER, F.: Untersuchungen über die Viskosität des Blutserums. Helv. physiol. pharmacol. Acta **8**, 486–504 (1950).

96. KROGH, A.: The anatomy and physiology of capillaries. New York: Hafner 1959.

97. LESSNER, A., ZAHIVE, J., SILBERBERG, A., FREI, E. H., DREYFUSS, F.: The viscoelastic properties of whole blood. In: Theoretical and clinical hemorheology. Proc. 2nd Internat. Conf. on Hemorheology, Heidelberg 1969. Berlin-Heidelberg-New York: Springer 1971.

98. LEVY, M., SHARE, R. L.: The influence of erythrocyte concentration upon the pressure flow relationship of the dog's hind limb. Circulat. Res. **1**, 247–255 (1953).

99. LEW, H. S., FUNG, Y. C.: Plug effect of erythrocytes in capillary blood vessels. Biophys. J. **10**, 80–90 (1970).

100. Lew, H. S., Fung, Y. C.: The motion of the plasma between the red cells in the bolus flow. Biorheology **6**, 109–119 (1969).

101. Lighthill, M. J.: Pressure-forcing of tightly fitting pellets along fluid-filled elastic tubes. J. Fluid Mech. **34**, 113–143 (1968).

102. Litton, A., Berman, H., Walters, C. W.: Quantification of the microvasculature of the hamster-cheekpouch. Anat. Rec. **154**, 472 (1966).

103. Mall, F.: Die Blut- und Lymphwege im Dünndarm des Hundes. Ber. sächs. Ges. Akad. Wiss. **14**, 151 (1888).

104. Mason, S. G.: The microrheology of suspensions with particular reference to blood. In: Theoretical and clinical hemorheology. Proc. 2nd Intern. Conf. on Hemorheology, Heidelberg 1969. Berlin-Heidelberg-New York: Springer 1971.

105. Meiselmann, H. J., Merrill, E. W., Gilliland, E. R., Pelletier, G. A., Salzman, E. W.: Influence of plasma osmolarity on the rheology of human blood. J. appl. Physiol. **22**, 772–781 (1967).

106. Mentzer, W. C., Baehner, R. L., Schmid-Schönbein, H., Robinson, S. H., Nathan, D. G.: Selective Reticulocyte destruction in erythrocyte pyruvate kinase deficiency. J. clin. Invest. (in press).

107. Merrill, E. W., Gilliland, E. R., Cokelet, G., Shin, H., Britten, A., Wells, R. E.: Rheology of blood and flow in the microcirculation. J. appl. Phys. **18**, 255–260 (1963).

108. — Cokelet, G. C., Britten, A., Wells, R. E.: Non-Newtonian rheology of human blood. Effect of fibrinogen deduced by substraction. Circulat. Res. **13**, 48–55 (1963).

109. — Gilliland, E. R., Cokelet, G., Shin, H., Britten, A., Wells, R. E.: Rheology of human blood, near and at zero flow. Effects of temperature and hematocrit level. Biophys. J. **3**, 199–233 (1963).

110. — Benis, A. M., Gilliland, E. R., Sherwood, T. K., Salzman, E. W.: Pressure flow relations of human blood in hollow fibers at low flow rates. J. appl. Physiol. **20**, 954–967 (1965).

111. — Rheology of human blood and some speculations on its role in vascular homeostasis. In: Biophysical mechanisms in vascular homeostasis (P. N. Sawyer, ed.), p. 121–137. New York 1965.

112. — Pelletier, G. A.: Viscosity of human blood: transition from Newtonian to non-Newtonian. J. appl. Physiol. **23**, 178–182 (1967).

113. — Rheology of blood. Physiol. Rev. **49**, 863–888 (1969).

114. Monro, P. A. G.: The appearance of cell-free plasma and "grouping" of red cells in normal circulation in small blood vessels observed in vivo. Biorheology **1**, 239–246 (1963).

115. — Progressive deformation of blood cells with increasing velocity of flowing blood. Bibl. anat. (Basel) **10**, 99–103 (1969).

116. Müller, A.: Abhandlungen zur Mechanik der Flüssigkeiten mit besonderer Berücksichtigung der Hämodynamik. Fascikel I: Die Newtonsche Strömung. Freiburg (Schweiz) u. Leipzig 1936.

117. Murphy, J. R.: Erythrocyte metabolism. VI. Cell shape and the location of cholesterol in the erythrocyte membrane. J. Lab. clin. Med. **65**, 756–774 (1965).

118. — The influence of pH and temperature on some physical properties of normal erythrocytes and erythrocytes from patients with hereditary spherocytosis. J. Lab. clin. Med. **69**, 758–775 (1967).

119. — Hemoglobin CC disease. Rheological properties of erythrocytes and abnormalities in cell water. J. clin. Invest. **47**, 1483–1495 (1968).

120. Neuschloss, S. M.: Die Viskosität des Blutes. Handbuch der normalen und pathologischen Physiologie, Bd. VI, Teil 1, S. 619–648. 1928.

121. Perutz, M. F., Liquori, A. M., Eirich, F.: X-Ray and solubility studies of the hemoglobin of sickle cell patients. Nature (Lond.) **167**, 929–931 (1951).

122. PLOMANN, H.: Ophtalmoskopischer Nachweis von Veränderung der Haltbarkeit von Blutkörperchenaufschwemmung. Hygiea (Stockh.) **82**, 363–373 (1920).

123. POISEUILLE, J. L. M.: Recherches expérimentales sur le mouvement des liquides de nature differente dans les tubes de trés petits diamètres. Ann. Chim. Physique **3**, 36 (1842).

124. DU PRÉ, A., DEMNING, P., WATSON, J. H.: The viscosity of the blood. Proc. roy. Soc. B **78** (1906).

125. PREUSSER, H. J.: Elektronenoptische Untersuchungen an Oberflächen von Membranfiltern. Kolloid-Z. **218**, 129–136 (1967).

126. PROTHERO, J. W., BURTON, A. C.: The physics of blood flow in capillaries. III. The pressure required to deform erythrocytes in acid-citrate-dextrose. Biophys. J. **2**, 213–222 (1962).

127. RAND, R. P., BURTON, A. C.: Mechanical properties of the red cell membrane. I. Membrane stiffness and intracellular pressure. Biophys. J. **4**, 115–135 (1964).

128. — Mechanical properties of the red cell membrane. II. Viscoelastic breakdown of the membrane. Biophys. J. **4**, 303–316 (1964).

129. RAND, P. W., LACOMBE, E.: Hemodilution, tonicity and blood viscosity. J. clin. Invest. **43**, 2214–2226 (1964).

130. — AUSTIN, W. H., LACOMBE, E., BAKER, N.: pH and blood viscosity. J. appl. Physiol. **25**, 550–559 (1968).

131. REINER, M., SCOTT-BLAIR, G. W.: Rheological terminology. In: Rheology, theory and application, vol. IV (F. R. EIRICH, ed.), p. 461–488. New York: Academic Press 1967.

132. ROSENBLUM, W. J.: Effects of Dextran 40 on blood viscosity in experimental macroglobulinaemia. Nature (Lond.) **218**, 591–593 (1968).

133. ROTHMAN, M.: Ist das Poiseuille'sche Gesetz für Suspensionen gültig? Pflügers Arch. ges. Physiol. **155**, 318–345 (1913).

134. ROWLANDS, S., GROOM, A. C., THOMAS, H. W.: The difference in circulation times between erythrocytes and plasma in vivo. Proc. 4th Congr. Rheology, vol. IV (A. L. COPLEY, ed.), p. 371–380. New York: Wiley 1964.

135. RUHENSTROHT-BAUER, G.: Mechanismus und Bedeutung der beschleunigten Erythrozytensenkung. Klin. Wschr. **44**, 531–539 (1966).

136. RUMSCHEIDT, F. D., MASON, S. G.: Particle motions in sheared suspensions. XI. Internal circulation in fluid drops (experimental). J. Colloid Sci. **16**, 210–237 (1961).

137. — — Particle motions in sheared suspensions. XII. Deformation and burst of fluid drops in shear and hyperbolic flow. J. Colloid Sci. **16**, 238–261 (1961).

138. SACKS, A. H., TICKNER, E. G.: Viscosity of blood. In: Theoretical and clinical hemorheology. Proc. 2nd Intern. Conf. Hemorheology, Heidelberg 1969. Berlin-Heidelberg-New York: Springer 1971

139. SCHMID-SCHÖNBEIN, H., GAETHGENS, P., HIRSCH, H.: Nicht-Newton'sche Viskosität des Blutes und Erythrozytenaggregation. Proc. III. Symp. Int. Anaest. Poznan 1967, p. 344–351.

140. — — — Eine neue Methode zur rheologischen Untersuchung von Erythrozytenaggregaten. Pflügers Arch. ges. Physiol. **297**, 107–112 (1967).

141. — — — On the shear rate dependence of red cell aggregation in vitro. J. clin. Invest. **47**, 1447–1454 (1968).

142. — WELLS, R. E., SCHILDKRAUT, R.: Miscroscopy and viscometry of blood flowing under uniform shear rate (rheoscopy). J. appl. Physiol. **26**, 674–678 (1969).

143. — — Rheological consequences of osmotic red cell crenation. Pflügers Arch. ges. Physiol. **307**, 59–69 (1969).

144. — — Fluid drop like transition of erythrocytes under shear. Science **165**, 288–291 (1969).

145. — — GOLDSTONE, J.: Influence of deformability of human red cells upon blood viscosity. Circulat. Res. **25**, 131–143 (1969).

146. Schmid-Schönbein, H., Wells, R. E.: Quantification of the dynamics of red cell aggregation. Bibl. anat. (Basel) **10**, 45–51 (1969).

146a. — — — Increased viscous resistance of blood due to hypertonicity: a possible mechanism for intrarenal distribution of blood flow. Fed. Proc. **28**, 716 abs. (1969).

147. Schmid-Schönbein, H., Goldstone, J., Wells, R. E.: Blood as an emulsion: red cell fluidity and high shear viscosity of red cell dispersions. Amer. J. Physiol. (in press).

148. — Hemorheological aspects of splenic function. In: Die Milz (K. Lennert u. D. Harms, eds.), p. 67–80. Berlin-Heidelberg-New York: Springer 1970.

149. — Goldstone, J., Wells, R. E.: Red cell deformation and red cell aggregation: their influence on blood rheology in health and disease. In: Theoretical and clinical hemorheology. Proc. 2nd Intern. Conf. Hemorheology Heidelberg 1969. Berlin-Heidelberg-New York: Springer 1971.

150. — Erythrocyte fluidity: Hemodynamic signification and methods of quantification in vitro. Proc. VI. Conf. on Microcirculation (Aalborg 1970). Bibl. anat. (Basel) **11** (in press).

151. Scott-Blair, G. W.: The importance of the sigma phenomenon in the study of the flow of blood. Rheol. Acta **1**, 123–126 (1958).

152. — An equation for the flow of blood, plasma and serum through glass capillaries. Nature (Lond.) **183**, 613–614 (1959).

153. Seaman, G. V. F., Swank, R. L.: The influence of electrokinetic charge and deformability of the red blood cell on the flow properties of its suspensions. Biorheology **4**, 47–59 (1967).

154. Skalak, R., Brånemark, P. I.: Deformation of red blood cells in capillaries. Science **164**, 717–719 (1969).

155. Skovborg, F., Nielsen, A. V., Schlichkrull, J., Ditzel, J.: Blood-viscosity in diabetic patients. Lancet **1966** I, 129–131.

156. Solvsteen, P., Kristjansen, P. J.: The effect of carbon dioxyde and oxygen on the viscosity of whole blood. Z. Kreisl.-Forsch. **57**, 42–46 (1968).

157. Somer, T.: The viscosity of blood, plasma and serum in dys- and paraproteinemias. Acta med. scand. **180**, Suppl. 456 (1966).

158. Stone, H. O., Thompson, H. K., Schmidt-Nielsen, K.: Influence of erythrocytes on blood viscosity. Amer. J. Physiol. **214**, 913–918 (1968).

159. Strumia, M. M., Phillips, M.: Effect of red cell factors on the relative viscosity of whole blood. Amer. J. clin. Path. **39**, 464–474 (1963).

160. Taylor, G. I.: The viscosity of a fluid containing small drops of another fluid. Proc. roy. Soc. A **138**, 41–44 (1932).

161. — The formation of emulsions in definable fields of low. Proc. roy. Soc. A **146**, 501–523 (1934).

162. Teitel, P.: Disk-sphere transformation and plasticity alteration of red blood cells. Nature (Lond.) **206**, 409–410 (1965).

163. — Pathophysiology of hemolytic anemias and microrheological competence of erythrocytes. Proc. XIII Congr. Hematology Munich 1970. (Biorheology 1971, in press).

163a: Tickner, E. G., Sacks, A. H.: Slow flow of rigid particles suspensions through simulated stenoses, Biorheology **5**, 275—283 (1968).

164. Thomas, D. G.: The transport characteristics of suspensions. VII. A note on the viscosity of Newtonian suspensions of uniform spherical particles. J. Coll. Sci. **20**, 267–277 (1965).

165. Thomas, H. W.: On the difference between the clearance curves of labelled red cells and labelled plasma from the circulatory bed of the heart and lung. Biorheology **3**, 36–40 (1965).

166. Thuransky, K.: Der Blutkreislauf der Netzhaut. Ung. Akad. Wiss. Budapest 1957.

167. THURSTON, G. B.: The viscoelasticity of blood and plasma during coagulation in circular tubes. Proc. VI. Conf. on Microcirculation, Aalborg 1970. Bibl. anat. (Basel) **11** (in press).

168. WAYLAND, H., MEISELMAN, H.: Viscometric measurements for blood and plasma. In: Theoretical and Clinical Hemorheology. Berlin-Heidelberg-New York: Springer 1971.

169. WEED, R. I., LaCELLE, P. L., MERRILL, E. W.: Metabolic dependence of red cell deformability. J. clin. Invest. **48**, 795 (1969).

170. WEIDENREICH, F.: Über die Form von Säugererythrozyten. Pflügers Arch. ges. Physiol. **132**, 143–147 (1910).

171. WELLS, R. E., Jr., DENTON, R., MERRILL, E. W.: Measurement of viscosity of biologic fluids by cone plate viscometer. J. Lab. clin. Med. **57**, 646–656 (1961).

172. — GAWRONSKI, T. H., COX, P. M., PERERA, R. D.: Influence of fibrinogen on flow properties of erythrocyte suspensions. Amer. J. Physiol. **207**, 1035–1040 (1964).

173. Blood flow in the microcirculation of man and the flow properties of blood. Bibl. anat. (Basel) **9**, 520–524 (1967).

174. — SCHMID-SCHÖNBEIN, H.: Red cell deformation and fluidity of concentrated red cell suspensions. J. appl. Physiol. **27**, 213–217 (1969).

175. — — GOLDSTONE, J.: Flow behavior of red cells in pathologic sera: existence of a yield shear stress in the absence of fibrinogen. In: Theoretical and clincal hemorheology. Proc. 2nd Int. Conf. Hemorheology, Heidelberg 1969. Berlin-Heidelberg-New York: Springer 1971.

176. WIEDEMAN, M. P.: Dimensions of blood vessels from the distributing artery to collecting vein. Circulat. Res. **12**, 375–378 (1963).

177. WILKINSON, W. L.: Non-Newtonian fluids: fluid mechanism, mixing and heat transfer. New York-London-Oxford-Paris: Pergamon Press 1960.

178. WHITTACKER, S. R. F., WINTON, F. R.: The apparent viscosity of blood flowing in the isolated hindlimb of the dog, and its variation with corpuscular concentration. J. Physiol. (Lond.) **78**, 339–369 (1933).

179. WHITMORE, R. L.: The interaction of forces in blood flow. Bibl. anat. (Basel) **9**, 240–245 (1967).

180. — A theory of blood flow in small vessels. J. appl. Physiol. **22**, 767–771 (1957).

181. — Rheology of the circulation. Oxford-London: Pergamon Press 1968.

182. ZIERLER, K. L.: Circulation times and the theory of indicator dilution methods for determining blood flow and volume. In: Handbook of physiology, sect. 2, Circulation, vol. I (W. F. HAMILTON, and P. DOW, eds.). Washington, D.C. 1962.

183. ZWEIFACH, B. W., RICHARDSON, D. R.: Pressure adjustments in the macro- and microvasculature of the perfused mesentery. Proc. VI. Conf. on Microcirculation (Aalborg 1970). Bibl. anat. (Basel) **11** (in press).

Namenverzeichnis

Die in Klammern stehenden Ziffern beziehen sich auf die Nummern der Zitate innerhalb des laufenden Textes und der Literatur.

Die gewöhnlich gesetzten Ziffern weisen auf die entsprechenden Stelle im Text und die kursiven Seitenzahlen auf das Literaturverzeichnis hin.

Sachverzeichnis

Ergebnisse der Physiologie

Biologischen Chemie und experimentellen Pharmakologie

Reviews of Physiology

Biochemistry and Experimental Pharmacology

Herausgeber / Editors

E. Helmreich, Würzburg · H. Holzer, Freiburg · R. Jung, Freiburg
K. Kramer, München · O. Krayer, Boston
F. Lynen, München · P. A. Miescher, Genf · W. D. M. Paton, Oxford
H. Rasmussen, Philadelphia · A. E. Renold, Genf
U. Trendelenburg, Würzburg · H. H. Weber, Heidelberg

Sonderdruck aus Band 63

H. O. Schild
Henry Hallett Dale, 1875—1968

Nicht im Handel

Springer-Verlag Berlin Heidelberg GmbH 1971

Inhalt

Ergebnisse der Physiologie

Biologischen Chemie und experimentellen Pharmakologie

Reviews of Physiology

Biochemistry and Experimental Pharmacology

Herausgeber / Editors

E. Helmreich, Würzburg · H. Holzer, Freiburg · R. Jung, Freiburg
K. Kramer, München · O. Krayer, Boston
F. Lynen, München · P. A. Miescher, Genf · W. D. M. Paton, Oxford
H. Rasmussen, Philadelphia · A. E. Renold, Genf
U. Trendelenburg, Würzburg · H. H. Weber, Heidelberg

Sonderdruck aus Band 63

R. F. Schmidt
Presynaptic Inhibition in the Vertebrate Central Nervous System
With 33 Figures

Nicht im Handel

Springer-Verlag Berlin Heidelberg GmbH 1971

Inhalt

Ergebnisse der Physiologie

Biologischen Chemie und experimentellen Pharmakologie

Reviews of Physiology

Biochemistry and Experimental Pharmacology

Herausgeber / Editors

E. Helmreich, Würzburg · H. Holzer, Freiburg · R. Jung, Freiburg
K. Kramer, München · O. Krayer, Boston
F. Lynen, München · P. A. Miescher, Genf · W. D. M. Paton, Oxford
H. Rasmussen, Philadelphia · A. E. Renold, Genf
U. Trendelenburg, Würzburg · H. H. Weber, Heidelberg

Sonderdruck aus Band 63

W. Schaper
The Physiology of the Collateral Circulation in the
Normal and Hypoxic Myocardium
With 10 Figures

Nicht im Handel

Springer-Verlag Berlin Heidelberg GmbH 1971

Inhalt

Ergebnisse der Physiologie

Biologischen Chemie und experimentellen Pharmakologie

Reviews of Physiology

Biochemistry and Experimental Pharmacology

Herausgeber / Editors

E. Helmreich, Würzburg · H. Holzer, Freiburg · R. Jung, Freiburg
K. Kramer, München · O. Krayer, Boston
F. Lynen, München · P. A. Miescher, Genf · W. D. M. Paton, Oxford
H. Rasmussen, Philadelphia · A. E. Renold, Genf
U. Trendelenburg, Würzburg · H. H. Weber, Heidelberg

Sonderdruck aus Band 63

H. Schmid-Schönbein and R. E. Wells, Jr.
Rheological Properties of Human Erythrocytes and their Influence
upon the "Anomalous" Viscosity of Blood

With 36 Figures

Springer-Verlag Berlin Heidelberg GmbH 1971

Inhalt

ATHERO-SCLEROSIS:

Proceedings of the Second International Symposium

Held in Chicago, Illinois, November 2nd — 5th, 1969

Sponsored by the
International Society of
Cardiology
(Scientific Council on Athero-
sclerosis and Ischaemic
Heart Disease
Scientific Council on Epidemio-
logy and Prevention)
European Atherosclerosis Group
American Heart Association
(Council on Arteriosclerosis
Council on Cerebrovascular
Disease
Council on Epidemiology)
Chicago Heart Association

Editorial committee:
L. Cohen, R. Eisenstein,
G. S. Getz, G. M. Hass,
H. J. McDonald, R. Pick,
A. Scanu, and J. Schweppe

Chairman: Richard J. Jones,
M. A., M. D.,
Associate Professor,
Department of Medicine,
Pritzker School of Medicine,
University of Chicago, IL, U.S.A.

With approx. 150 figures
XXXII, 706 pages. 1970
Cloth DM 65,—; US $ 18.00

This symposium was planned to provide an up-to-date review of the pathogenetic processes involved in atherosclerosis and of proposals that have been made for their control. It includes sections on the basic pathogenesis of atherosclerosis, arterial thrombosis and their resulting complications; epidemiological considerations in coronary heart disease from around the world; nutritional studies relating to atherosclerosis; newer knowledge about the serum lipoproteins and the regulation of lipid metabolism, both sterols and triglycerides; considerations of prophylactic as well as drug therapy directed against pathogenetic mechanisms; and finally a discussion of progress to date and future prospects of plans directed toward the control of atherosclerosis.

Contents

Pathogenesis of Atherosclerosis. The Reactions of the Arterial Wall. Thrombosis and Atherosclerosis. Selected Papers on Pathogenesis of Atherosclerosis, Including Thrombosis. Serum Lipoproteins. Selected Papers on Lipoproteins and Atherosclerosis. Regulation of Triglycerides, Including Carbohydrate-Lipid Interaction. Sterol Balance and Metabolism. Selected Papers on Lipid Metabolism. Environmental and Host Factors in Coronary Heart Disease, Including Risk Factors: An Epidemiological View. Selected Papers on Epidemiology of Atherosclerosis. Nutritional Studies and Atherosclerosis. Selected Papers on Nutritional Studies. Panel Discussion on Pathogenesis as it may Influence Prevention and Therapy. Recent Advances in Drugs Affecting Lipids, Platelets and Autonomic Nerve Mediators. Selected Papers on Drug Effects. Progress in the Control of Atherosclerosis. Program Planning for Control of Atherosclerosis. Summary of Symposium.

■ **Prospectus on request**

SPRINGER-VERLAG
BERLIN · HEIDELBERG · NEW YORK

Universitätsdruckerei H. Stürtz AG, Würzburg